The Radiation Legacy of the Soviet Nuclear Complex

The International Institute for Applied Systems Analysis

is an interdisciplinary, nongovernmental research institution founded in 1972 by leading scientific organizations in 12 countries. Situated near Vienna, in the center of Europe, IIASA has been for more than two decades producing valuable scientific research on economic, technological, and environmental issues.

IIASA was one of the first international institutes to systematically study global issues of environment, technology, and development. IIASA's Governing Council states that the Institute's goal is: *to conduct international and interdisciplinary scientific studies to provide timely and relevant information and options, addressing critical issues of global environmental, economic, and social change, for the benefit of the public, the scientific community, and national and international institutions*. Research is organized around three central themes:

- Environment and Natural Resources;
- Energy and Technology;
- Population and Society.

The Institute now has national member organizations in the following countries:

Austria
The Austrian Academy of Sciences

Bulgaria*
The Bulgarian Committee for IIASA

Finland
The Finnish Committee for IIASA

Germany**
The Association for the Advancement of IIASA

Hungary
The Hungarian Committee for Applied Systems Analysis

Japan
The Japan Committee for IIASA

Kazakhstan*
The Ministry of Science –
The Academy of Sciences

Netherlands
The Netherlands Organization for Scientific Research (NWO)

Norway
The Research Council of Norway

Poland
The Polish Academy of Sciences

Russian Federation
The Russian Academy of Sciences

Slovak Republic*
The Slovak Committee for IIASA

Sweden
The Swedish Council for Planning and Coordination of Research (FRN)

Ukraine*
The Ukrainian Academy of Sciences

United States of America
The US Committee for IIASA

*Associate member
**Affiliate

The Radiation Legacy of the Soviet Nuclear Complex

An Analytical Overview

Nikolai N. Egorov, Vladimir M. Novikov, Frank L. Parker, and Victor K. Popov, Editors

IIASA

earthscan from Routledge

First published in the UK in 2000 by
Earthscan Publications Ltd

This edition published 2013 by Earthscan

For a full list of publications please contact:

Earthscan
2 Park Square, Milton Park, Abingdon, Oxfordshire OX14 4RN
Simultaneously published in the USA and Canada by Earthscan
711 Third Avenue, New York, NY 10017

First issued in paperback 2016

Earthscan is an imprint of the Taylor & Francis Group, an informa business

A catalogue record for this book is available from the British Library

ISBN 13: 978-1-138-98446-2 (pbk)
ISBN 13: 978-1-85383-658-9 (hbk)

Typesetting by IIASA

Cover design by Yvonne Booth

Contents

List of Figures

List of Tables

List of Contributors

A.A. Burykin
All-Russia Research Institute of Chemical Technology (VNIICHT), Moscow, Russia (Chapter 1)
A.A. Iskra
VNIICHT, Moscow, Russia (Chapter 1)
A.A. Vlasov
Ministry of the Russian Federation for Atomic Energy (Minatom), Moscow, Russia (Chapter 2)
R.A. Svetlakova
Minatom, Moscow, Russia (Chapter 2)
V.I. Zemlyanukhin
VNIICHT, Moscow, Russia (Chapter 3)
V.N. Rud'
Joint Stock Society (AO) TVEL, Moscow, Russia (Chapter 3)
B.A. Bezrukov
Concern Rosenergoatom, State Enterprise, Moscow, Russia (Chapter 4)
E.D. Stukin
Institute of Global Climate and Ecology (IGKE), Moscow, Russia (Chapter 4)
L.P. Kham'yanov
All-Russia Research Institute for Atomic Power Plants (VNIIAES), Moscow, Russia (Chapter 4)
S.B. Khubetsov
VNIIAES, Moscow, Russia (Chapter 4)
O.E. Kiknadze
Russian Research Center–Kurchatov Institute (RRC–KI), Moscow, Russia (Chapter 5)
Yu.V. Sivintsev
RRC–KI, Moscow, Russia (Chapter 5)

N.V. Arkhangelskii
Minatom, Moscow, Russia (Chapter 6)
P.M. Egorenkov
RRC–KI, Moscow, Russia (Chapter 6)
A.I. Rybalchenko
All-Russia Research and Design Institute of Production Engineering (VNIPIPT), Moscow, Russia (Chapter 7)
I.S. Serebryakov
VNIICHT, Moscow, Russia (Chapter 7)
B.V. Gorobets
Minatom, Moscow, Russia (Chapter 8)
D.F. Dul'nev
Minatom, Moscow, Russia (Chapter 8)
A.M. Matushchenko
Minatom, Moscow, Russia (Chapter 9)
E.D. Stukin
IGKE, Moscow, Russia (Chapter 9)
A.I. Sobolev
Moscow Scientific and Industrial Association (SIA) Radon (Chapter 10)
V.A. Tikhomirov
Moscow SIA Radon (Chapter 10)

Editorial Group
Nikolai N. Egorov
Vladimir M. Novikov
Frank L. Parker
Victor K. Popov

Editor-in-Chief
Nikolai N. Egorov
Deputy Minister, Ministry of the Russian Federation for Atomic Energy

Acknowledgments

The contributors and editorial group would like to thank the U.S. Department of Energy for providing the grant for this International Institute for Applied Systems Analysis (IIASA) study and for assistance to the Radiation Legacy (Radleg) project, and the European Union for financial contribution to the Radleg project through the International Science and Technology Center (ISTC) in Moscow.

We appreciate the input of Academician N.P. Laverov (Vice President of the Russian Academy of Science), Academician E.P. Velikhov (President of the Russian Research Center–Kurchatov Institute [RCC–KI]), and Dr. P. de Jánosi (former Director of IIASA). These individuals were instrumental in the initiation of both the IIASA and ISTC projects.

We are also thankful to:

- Professor G. MacDonald (IIASA Director), Academician N.N. Ponomarev-Stepnoy (Vice President of RRC–KI), Professor D.I. Skorovarov (Director of the All-Russia Research Institute of Chemical Technology [VNIICHT]), Professor V.V. Shatalov (Deputy Director of VNIICHT), and Academician Yu.A. Israel (Director of the Institute of Global Climate and Ecology [IGKE]) for their steady support.
- B.K. Bylkin, Yu.E. Gorlinskii, N.E. Kukharkin, and O.G. Lebedev (RRC–KI); A.V. Komarov (VNIICHT); and B.A. Reznik (Rosenergoatom) for their helpful consultations.
- R.S. Gudimov, I.G. Nekhorosheva, N.M. Sergeeva, and V.G. Yurovskikh (RRC–KI); Yu.B. Mostinskii and V.V. Semchenko (VNIICHT); and U. Neudeck and K. Novikova (IIASA) for providing technical and organizational support.

Preface

Vladimir M. Novikov and Frank L. Parker

Over the second half of the 20th century, vast quantities of radioactive waste (RW) have accumulated in numerous radioactively contaminated sites throughout the world, particularly in the United States and Russia. These wastes and contaminated sites were the result of the production of nuclear weapons, nuclear accidents, and practices in the early days of nuclear industry, as well as the routine and normal procedures of nuclear power plants. Once released to the environment, radioactive material (RM) can become a significant transboundary problem and cause international concern about its impact on health and the environment.

The end of the Cold War made possible the removal of the curtain that shielded the nuclear industry from the scrutiny of public and ecological organizations and opened the way for a systematic, comprehensive, and interdisciplinary international effort to study the problems of nuclear inheritance of the past half century. Such a study would facilitate decision making to achieve technically sound, publicly acceptable, and cost-effective policies oriented to remediation and/or stabilization of the burden of the radiation inheritance.

In response to this challenge, the International Institute for Applied Systems Analysis (IIASA) has established the Radiation Safety of the Biosphere (RAD) project to conduct a series of studies on these problems. The Advisory Committee of the RAD project (cochaired by Dr. J. Ahearne and Academician N. Laverov) recommended starting with the problems of the nuclear inheritance of the former Soviet Union (FSU), where the most severe radiological problems are located. The RW problems there are complicated by the stresses of rapid social, economic, and political changes in the region.

The effort to collect and verify data that will provide an overview of the radiation contamination of the FSU and serve as inputs to risk assessment and other applications, such as transboundary risk management and radionuclide material flows, is a major task. No single institution can collect and collate all the data relevant to the nuclear inheritance of the FSU. In addition, these data, as well as statistical data on the generation, content, and distribution of the RM are partially published in some documents for restricted or in-house use only and partially stored in archives that have been declassified. Therefore, the decision was taken to encourage the International Science and Technology Center (ISTC) in Moscow to start a project to create a network of scholars and professionals at participating institutions in Russia with the goal of developing what became known as the "Data System for Evaluation of the Radiation Legacy of the FSU."

This approach was approved and encouraged by Dr. N. Egorov, Deputy Minister of the Ministry of the Russian Federation for Atomic Energy (Minatom), and Academician E. Velikhov, President of the Russian Research Center–Kurchatov Institute (RRC–KI). IIASA helped to develop the detailed Russian proposal to the ISTC and provided methodological support for the Russian studies. In 1995, the Russian project started under the name ISTC #245 Radleg (for *rad*iation *leg*acy), and IIASA became its outside collaborator. It was decided to start with preparation of the sector overview that would review information about RW accumulated and plans for nuclear material management available in organizations participating in the Radleg project (about 20 of the most important organizations). *Figure P.1* illustrates the structure of Radleg and its collaboration with the IIASA RAD project.

It was recommended that descriptions of the following sectors should be prepared:

- Mining and processing of uranium ores.
- Uranium conversion and isotopic enrichment.
- Nuclear fuel fabrication.
- Nuclear power plants.
- Nuclear propulsion power installations, including the nuclear ships and vessels of the navy and civilian nuclear-powered ships.
- Nuclear research reactors and nuclear research centers.

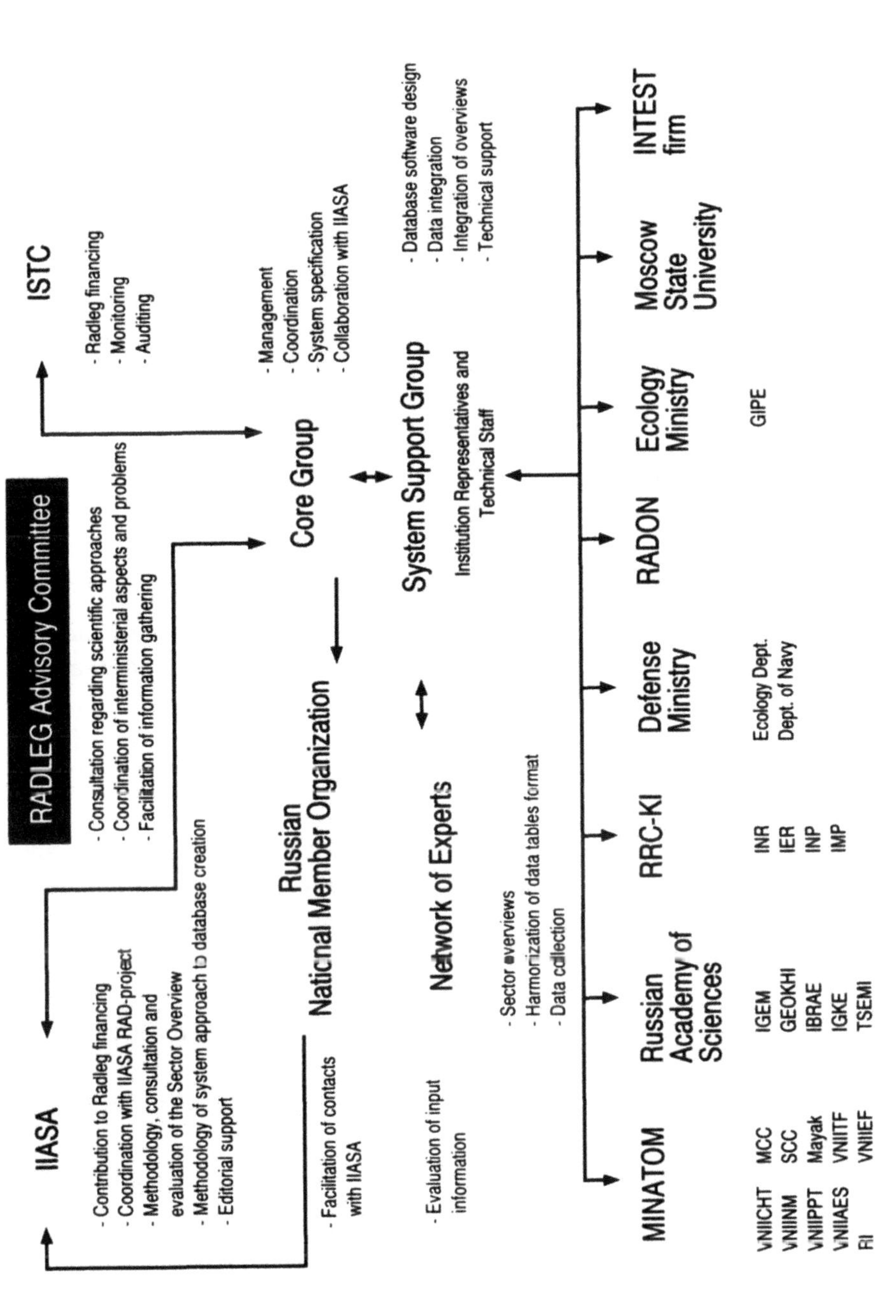

Figure P.1. Cooperation of the IIASA RAD project with Russian institutions.

- Plutonium production and radiochemical reprocessing of spent nuclear fuel.
- Manufacturing and utilization of nuclear warheads.
- Nuclear explosions, including nuclear weapons tests and nuclear explosions for civilian purposes.
- Storage and processing of nonreactor RWs and sources of ionizing radiation.

In each sector, the RAD project, jointly with Radleg, formulated requirements for the material pertinent to each sector of the nuclear complex of the FSU.

Each chapter attempts to address the following topics:

1. General information.

    1.1. Brief characteristics and functional role of the sector.
    1.2. Structure of the sector (what enterprises make up the sector).
    1.3. Geographical location and ministerial subordination of the sector's enterprises.
    1.4. Production capacities and operation characteristics of the enterprises.

2. Characteristics of accumulated RM and RW for the sector, total, and by individual enterprises as of 1 January 1996. If the data of 1 January 1996 are absent, an explanation of why is given and the most relevant available data are presented.

    2.1. Amount of accumulated RM and RW (volume and activity).
    2.2. Nuclide composition (if available).
    2.3. Aggregation state (liquid, solid) and specific activity levels (low, medium, or high).
    2.4. Releases and discharges of radionuclides into the environment (absolute and in percentage of the permissible levels).[1]
    2.5. General characteristics of RM and RW storage.

[1]It was agreed to use current Russian radiation protection standards, which are based on the recommendations of the International Commission on Radiological Protection (ICRP) and summarized in *Norms of Radiation Safety* (NRB–06), Goskomsanepidnadzor of the Russian Federation, Moscow, 1996.

2.5.1. Storage type.
2.5.2. Capacity (total and percentage filled).
2.5.3. Storage data.
2.5.4. Storage location.

2.6. Territories contaminated by radionuclides as a result of the enterprise's operation.

2.6.1. Accidents in the sector's enterprises resulting in radioactive contamination of the environment.
2.6.2. Contamination of sanitary-protection zone and territory outside the zone.
2.6.3. General data on contamination (area, specific activity).

To facilitate the preparation of the sector overview, the IIASA RAD project provided additional support for the work in Russia and for its own studies on Chelyabinsk–65, Tomsk–7, and Krasnoyarsk–26.

The first version of the sector overview was drafted in 1996. This book presents a new version of the sector overview that was examined by internationally recognized reviewers. Editing of the manuscript was completed in 1998, and we hope it will serve as a stand-alone document.

In 1997, Battelle Press published *Behind the Nuclear Curtain: Radioactive Waste Management in the Former Soviet Union*, by Don J. Bradley, who has been following the Russian RW problems for many years. His book contains numerous other data and offers a rather detailed treatment of the nuclear fuel cycle as well as site analyses. Some of the summary tables in Bradley's book are taken from papers presented by the Russians at an International Atomic Energy Agency (IAEA) meeting in 1995; there were some discrepancies in the material presented at that meeting.

The data in this book were prepared de novo by the ministries that presented the data in 1995 and were examined by the Radleg expert group from Minatom. The data are more recent, have come from a common source, and have Minatom's imprimatur. Thus, this document is the only book that has received official approval from key Russian nuclear organizations that have to cope with the nuclear inheritance of the FSU.

The overview may be considered a basis for systematizing the information in a unified methodology. Particular attention is paid to those conditions that present a greater radiation hazard for the biosphere. It should be noted that this report is an analytical overview. The purpose is to provide the data to allow analysis of the consequences of the generation, storage, and releases of RM. The data presented here allow others to do similar analyses at other sites.

An Introduction to the Nuclear Complex of the Former Soviet Union

Nikolai N. Egorov

Radioactive contamination of the biosphere has occurred over the past 50 years due to nuclear weapons production and testing, enterprises of nuclear power engineering, military and civilian nuclear-powered ships, and nuclear explosions. Today, this contamination is one of the most serious global environmental problems.

The "nuclear club" countries – particularly the former Soviet Union (FSU) and the United States (USA) – have been the main contributors to the radioactive contamination of the biosphere. In the FSU, a significant amount of radioactive substances has entered into the environment due to nuclear weapons production and testing, large-scale radiation accidents (Kyshtym, 1957; Chernobyl, 1986; etc.), dumping of liquid radioactive wastes (LRWs), and disposal of solid radioactive wastes (SRWs) into seas. Some areas of the FSU are contaminated by radionuclides beyond permissible standards, and therefore some remediation measures must be applied. Radioactive wastes (RWs) and spent nuclear materials, which have accumulated as a result of past and present activities, are being managed in a way that does not completely meet modern international standards of radiation safety, and this poses a potential hazard for the biosphere.

Simultaneously, the termination of the Cold War facilitated significant openness of data concerning these problems. This information caused well-grounded anxiety among the public in Russia and other countries in the world. The anxiety was caused by a lack of nuclear safety in Russia and in other members of the Commonwealth of Independent States (CIS), and the possibility of transboundary transfer of radioactive contaminants to neighboring states and into the biosphere.

Keeping in mind the broadening of international cooperation in the effort to overcome radiation consequences of the Cold War, it seems important to have the most complete and trustworthy knowledge of the present state of radioactive hazardous objects in the territory of the FSU, particularly Russia. This information could help in setting priorities for RW management and for international cooperation in this field.

The evaluation of the radiation legacy of the FSU, which is an extremely urgent task, should be performed with the help of modern information technologies for collection, storage, and processing of data. The analysis should study the presence, origin, physical and chemical characteristics, and storage, treatment, and disposal of RWs and materials in the territories of Russia and other CIS members.

In Russia, problems of the radiation legacy are being handled in the framework of the Russian Federal Program "Management of Radioactive Wastes and Spent Nuclear Materials, Their Utilization and Disposal for 1996–2005." An important contribution to this program is the International Science and Technology Center (ISTC) Radleg project. The International Institute for Applied Systems Analysis (IIASA) is an initiator of this project and is its major customer. The results of the Radleg project were to provide major inputs into the broader Radiation Safety of the Biosphere (RAD) project, being developed by IIASA.

Historically, the nuclear complex in the Soviet Union was created to produce nuclear and thermonuclear weapons. After 1954 (startup of the world's first nuclear power plant in Obninsk), active use of nuclear energy for civilian power production as well as for ship propulsion (in both nuclear submarines and surface ships, including icebreakers) began in the country. Later, underground nuclear explosions were used to perform various national economy tasks.

The structure of the nuclear complex of the FSU included:

- *The nuclear weapon production line* – plutonium and tritium production reactors, nuclear fuel manufacturing for the reactors, highly enriched uranium (HEU) production, the production reactors' spent

nuclear fuel (SNF) reprocessing aimed at plutonium recovery, nuclear weapon components production from metallic HEU and plutonium, and plants and institutions engaged in design and manufacturing of nuclear warheads and related devices. The production works for manufacturing nuclear fuel for naval ship propulsion reactors (SPRs) and facilities for SNF reprocessing may be also attributed to this line.
- *The nuclear power production line* – nuclear power plants, research reactors, civilian nuclear SPRs, nuclear fuel manufacturing plants and SNF reprocessing plants, facilities for production of radioactive isotopes and ionizing radiation sources for use in the national economy, and enterprises for nonreactor RW processing and disposal (Radon Special Enterprises).

The production of primary nuclear materials for both military and civilian purposes was conducted, as a rule, at common industrial sites, but in different process facilities. These materials were obtained by processes such as uranium ore mining and enrichment, uranium isotopic enrichment, UF_6 conversion, nuclear fuel manufacturing, radiochemical reprocessing of SNF, and RWs processing.

These special industrial sites were chosen for interconnected production works because of their previous military production experience, geographic location, and availability of power, water, and labor. On the basis of these criteria such well-known enterprises as the Industrial Association (IA) Mayak in Chelyabinsk region, the Siberian Chemical Combine (SCC) in Tomsk region, and the Mining & Chemical Combine (MCC) in Krasnoyarsk land were created.

As a rule, the enterprises are collectively responsible for the complete technological cycle, which includes the management of production nuclear reactors, the reactors' SNF reprocessing plants, facilities for manufacturing necessary elements from weapons-grade uranium and plutonium, facilities for processing RWs, and auxiliary production complexes.

Later, new production works were created at the enterprises. Storage and reprocessing of SNF from civilian nuclear power plants were located at the MCC and IA Mayak. The production of radioactive isotopes and ionizing radiation sources used in the national economy were located at IA Mayak and the SCC. Thus, the enterprises are being transformed into powerful industrial complexes with tens of thousands of employees.

The nuclear complex of the FSU was vertically integrated. Everything necessary for its functioning was produced either at the enterprises and in organizations belonging to the Ministry of Medium Machine Building or in cooperation with enterprises and organizations of other ministries. These facilities were located in the different republics of the Soviet Union. Today, the ministry responsible for these works in Russia is the Ministry of the Russian Federation for Atomic Energy (Minatom).

Table I.1 presents the distribution of the nuclear industrial potential of the FSU, including countries of the CIS and Baltic region, formed after the Soviet Union's disintegration in 1991. Russia has inherited more than 80% of the nuclear industrial potential of the FSU. In 1996, removal of nuclear weapons from Ukraine, Belorussia, and Kazakhstan was completed. At present, all the nuclear weapons of the FSU reside in Russia.

In addition, following the FSU's disintegration, the construction of the fourth power unit at the Balakovo nuclear power plant (NPP) in Russia and the third power unit at the South Ukrainian NPP was completed. Data on the location of the FSU nuclear complex infrastructure in regions of the Russian Federation are presented in *Table I.2*. Similar data for other states, formed as a result of the FSU's disintegration, are presented in *Table I.3*. Data in both tables show that production of all nuclear fissile materials for military and civilian purposes in the FSU was concentrated in Russia. The production works are the main source of RW generation and accumulation.

In *Table I.4*, data on total amounts of RWs and SNF accumulated in Russia are presented, as of 1 January 1996. The data show that, during the time period of the FSU nuclear complex activities, more than 600 million m^3 of RWs with $\sim$1.9 billion Ci total activity and 8,700 tons of SNF with $\sim$5 billion Ci total activity have accumulated on the territory of Russia.

Spent nuclear fuel is stored in special structures (steel reservoirs, concrete and reinforced concrete storage facilities, etc.) and is kept under constant guard.

Figure I.1 presents the structure of the state administration of RW management in Russia, as of 1 January 1996.

The FSU nuclear complex formation and development, especially during the initial period, was accompanied by negative effects on the environment. This was due to production activities that discharged LRWs into the Techa River and later into the Karachay Lake and other open water

Table I.1. Distribution of nuclear potential as of 1991 in new independent states of the FSU after the disintegration of the USSR.

	Russia	Ukraine	Belorussia	Kazakhstan	Central Asia	TransCaucasus	Baltic region
Nuclear weapons	[a]	[a]	[a]	[a]	–	–	–
Nuclear ships	100%	–	–	–	–	–	–
NPPs	28 Power units, 20.2 GWe	14 Power units, 12.9 GWe	–	1 Power units, 0.3 GWe	–	2 Power units, 0.8 GWe	2 Power units, 3.0 GWe
Research reactors	25	1	1	2	1	1	1
Uranium ore mining & processing	40%	20%	–	10%	30%	–	–
Fuel elements manufacturing	100%	–	–	–	–	–	–
UO_2-pellets production	10%	–	–	90%	–	–	–
Uranium enrichment	100%	–	–	–	–	–	–
SNF reprocessing & RW treatment	100 %	–	–	–	–	–	–

[a]Indicates possession of nuclear weapons (data enumerating totals among these states is not available).

Table I.2. Nuclear complex infrastructure of the FSU inherited by the members of the RF.[a]

Regions	Nuclear objects, radioactive material storage facilities
Arkhangelsk region	SPRs of submarines and surface ships, enterprises for repair and utilization of nuclear submarines (Severodvinsk).
Republic of Bashkortostan	Enterprise for nonreactor RW management (Radon, Ufa).
Volgograd region	Enterprise for nonreactor RW management (Radon).
Voronezh region	The Novovoronezh NPP: five reactors (VVER–265; VVER–310; VVER–440, two; VVER–1000). The first two reactors have been decommissioned.
Irkutsk region	UF_6 production, uranium enrichment (Angarsk), enterprise for nonreactor RW management (Radon).
Kaluga region	State Research Center Institute of Physics and Power Engineering (Obninsk): research reactors and critical assemblies, nuclear materials, and RW storage facilities.
Kamchatka region	Navy SPRs.
Kirov region	Processing of uranium ore concentrates (Kirovo–Chepetsk).
Krasnoyarsk land	Uranium enrichment (Divnoguzsk), plutonium production reactors (two of the three reactors have been shut down), reprocessing of the production reactors' SNF, storehouse of SNF of VVER–1000 reactors, RT–2 plant for reprocessing of SNF of VVER–1000 reactors (in construction), RW storage and underground disposal (Zheleznogorsk), research reactor (Norilsk).
Kursk region	The Kursk NPP: four RBMK–1000-type reactors, storehouse of SNF.
Leningrad region & St. Petersburg	The Leningrad NPP: four RBMK–1000-type reactors, storehouse of SNF. Radon enterprise for nonreactor RW management (Sosnovyi Bor). The Institute of Nuclear Physics: research reactor (Gatchina).
Moscow region & Moscow city	Nuclear fuel element manufacturing for NPPs and Navy (Elektrostal), Russian Research Center–Kurchatov Institute: research reactors. Moscow State Institute of Engineering Physics: research reactor. The Institute for Theoretical and Experimental Physics: a research reactor. Moscow Scientific & Industrial Association Radon enterprise (Moscow). The Joint Institute for Nuclear Research: research reactors (Dubna).
Murmansk region	The Kola NPP: four VVER–440-type reactors, the SNF storehouse (Polyarnye Zori). Navy SPRs, plants for repair and utilization of nuclear submarines, the base of nuclear icebreakers, enterprises for processing and storage of RW and SNF of the Navy NPPs. Radon enterprise for nonreactor RW management.
Nenets autonomous district	Nuclear test site (the Novaya Zemlya Archipelago).
Nizhniy Novgorod region	Enterprises for nuclear weapons design and production (Sarov). Radon enterprise for nonreactor RW management.

[a]Enterprises for nonreactor RW management are listed in *Table 10.3*.

Table I.2. Continued.

Regions	Nuclear objects, radioactive material storage facilities
Novosibirsk region	Novosibirsk Plant of Chemical Concentrates: metallic uranium and uranium oxides production, manufacturing of nuclear fuel.
Penza region	Nuclear warheads production (Zarechny).
Primorsk land	Navy SPRs, storehouses of RW and SNF of Navy nuclear reactors, plants for repair and utilization of nuclear submarines (Bolshoi Kamen).
Rostov region	The Rostov NPP (in construction). Radon enterprise for nonreactor RW management.
Samara region	Radon enterprise for nonreactor RW management.
Saratov region	The Balakovo NPP: four VVER–1000-type reactors, a storehouse of SNF and RW. Radon enterprise for nonreactor RW management.
Smolensk region	The Smolensk NPP: three RBMK–1000-type reactors, a storehouse of SNF and RW (Desnogorsk).
Stavropol region	Uranium ore mining (ceased).
Sverdlovsk region	Production, storage, and dismantling of nuclear warheads (Lesnoy). Uranium enrichment (Novouralsk). The Byeloyarsk NPP: BN–600-type reactor, a storehouse of SNF. A research reactor (Ekaterinburg). Radon enterprise for nonreactor RW management.
Republic of Tatarstan	Radon enterprise for nonreactor RW management.
Tver region	The Kalinin NPP: two VVER–1000-type reactors, a storehouse of SNF and RW (Udomlya).
Tomsk region	UF_6 production and uranium enrichment, manufacturing of nuclear weapons components, plutonium production reactors (three of the five reactors have been shut down), reprocessing of the production reactors' SNF, RW storage and underground disposal (Seversk).
Ulyanovsk region	State Research Center–The Institute of Atomic Reactors: research reactors, MOX fuel manufacturing, underground disposal of liquid RW (Dimitrovgrad).
Udmurt Republic	Uranium ore processing, production of UF_4 and metallic uranium, manufacturing of nuclear fuel for production reactors (Glazov).
Khabarovsk land	Radon enterprise for nonreactor RW management.
Chelyabinsk region	Nuclear weapons' design (Snezhinsk), production and dismantling of nuclear warheads (Trekhgorny), plutonium production reactors (all five reactors have been shut down); reprocessing of SNF of marine reactors, research reactors, and nuclear power VVER–440, BN–350, and BN–660-type reactors; manufacturing of nuclear weapons' components, RW processing (Ozersk).
Chechen Republic	Radon enterprise for nonreactor RW management.
Chita region	Uranium ore mining and processing (Krasnokamensk).
Chukotsk autonomous district	The Bilibino NPP: four EGP–6-type reactors, a storehouse of SNF and RW.

reservoirs at the IA Mayak in Chelyabinsk region. It was also due to releases of cooling waters of plutonium production reactors into the Yenisei River at the MCC near Krasnoyarsk and into the Tom River at the SCC near Tomsk. Finally, the contamination was also due to accidents and atmospheric nuclear weapons tests and some civilian underground nuclear explosions. The accidents included the explosion of a tank with LRWs in 1957 at the IA Mayak, Chelyabinsk region, the wind-borne scattering of radioactive dust from the banks of the Karachay Lake in 1967 at the IA Mayak in Chelyabinsk region, and the Chernobyl accident in 1986. The accidents have led to radioactive contamination of large areas.

In the following chapters, an analytical overview of the detailed data on the radiation legacy of the FSU are presented for each sector of the nuclear complex. Given the complexity of the problem and in some cases the inaccessibility of important data, as well as the inconsistency of data published by different sources, it was not possible to present adequate information (either in volume or in content) in all the individual reviews.

Numeric data related to objects located in the territory of the Russian Federation are given mainly as of January 1997 or January 1996. For enterprises in other independent states of the FSU, data from January 1991 or January 1990 are used because data for later periods were officially unavailable.

Table I.3. Nuclear complex infrastructure of the FSU accommodated by states other than the RF, as a result of the disintegration of the USSR.

States	Nuclear objects, radioactive material storage facilities
Armenia	The Armenian NPP: two VVER–440-type reactors, storehouse of SNF and RW (Saskochevan) Enterprise for nonreactor RW management (Yerevan Special Combine).
Belorussia	The Institute of Power Engineering Problems: research reactor, critical assemblies (Sosny). Enterprise for nonreactor RW management (Minsk SE for RW Management).
Estonia	The Nuclear Submarines Training Center (Paldiski). Enterprise for nuclear ore processing – closed (Sillamae).
Georgia	The Institute of Physics: research reactor (Tbilisi). Enterprise for nonreactor RW management (Tbilisi).
Kazakhstan	Uranium ore mining, enrichment, and processing (Stepnogorsk, Aktau). Production of UO_2 pellets (Ust–Kamenogorsk). The Shevchenko NPP: BN–350-type reactor, storehouse of SNF and RW (Aktau). Enterprise for nonreactor RW management (Alma-Ata).
Kyrgystan	Uranium ore mining, enrichment, and processing (Mailisay, Mingush, Kara-Balta). Enterprise for nonreactor RW management (Kyrgystan Special Combine).
Latvia	The Institute of Physics: research reactor (Riga). Enterprise for nonreactor RW management (Riga).
Lithuania	The Ignalina NPP: two RBMK–1500-type reactors, storehouse of SNF and RW (Visaginas). Enterprise for nonreactor RW management (Special Combine at the Institute of Physics of the Academy of Sciences of Lithuania.
Moldavia	Enterprise for nonreactor RW management (Kishinev).
Tajikistan	Uranium ore mining, enrichment, and processing (Khodzhent, Gafurov, Tabashary, Charkassar). Enterprise for nonreactor RW management (Dushanbe).
Turkmenia	Enterprise for nonreactor RW management (Ashkhabad).
Ukraine	Uranium ore mining, enrichment, and processing (Zhyoltye Vody, Kirovograd). Uranium ore processing (Dnyeprodzerzhinsk). The Chernobyl NPP: three RBMK–1000-type reactors, storehouse of SNF and RW (Pripyat). The Rovno NPP: two VVER–440-type reactors, one VVER–1000-type reactor, storehouse of SNF and RW (Kuznetsovsk). The Khmelnitsky NPP: VVER–1000-type reactor, storehouse of SNF and RW (Slavuta). The South Ukrainian NPP: three VVER–1000-type reactors, storehouse of SNF and RW (Yuzhnoukrainsk). The Zaporozhe NPP: five VVER–1000-type reactors, storehouse of SNF and RW (Energodar). The Institute for Nuclear Research: research reactor (Kiev). Enterprises for nonreactor RW management (Kiev, Kharkov, Odessa, Donetsk, Lvov, Dnepropetrobsk).
Uzbekistan	Uranium ore mining, enrichment, and processing (Uchkuduk, Nurabad, Zarafshan, Navoi). The Institute of Nuclear Physics: research reactor (Tashkent). The Photon Corporation: research reactor (Tashkent) Enterprise for nonreactor RW management (Ulugbek).

Table I.4. Amounts of accumulated RW and SNF in the RF.

Ministries, departments, and organizations	LRW		SRW		SNF	
	m^3	Ci	m^3	Ci	Tons	Ci
Ministry of the Russian Federation for Atomic Energy (Minatom) Uranium ore mining and processing, uranium enrichment, nuclear fuel manufacturing, nuclear power production at NPPs, SNF reprocessing and nuclear weapon materials production	4.0×10^8	1.7×10^9	2.2×10^8	2.2×10^8	8,700	4.6×10^9
Ministry of Defense of the Russian Federation (Navy) Operation and utilization of nuclear ships and submarines	1.4×10^4	1.2×10^2	1.3×10^4	8.0×10^2	30	1.5×10^7
Ministry of Economy of the Russian Federation Department of Defense Industry Construction, repair, and utilization of nuclear ships and submarines	3.2×10^3	5.0	1.5×10^3	1.0×10^2	–	–
Ministry of Transport of the Russian Federation Operation and utilization of nuclear icebreakers	4.4×10^2	1.5×10^3	7.3×10^2	1.0×10^6	10	4.7×10^7
Radon Special Enterprises Processing and disposal of radioactive materials, used in medicine, scientific research, industry, etc.	–	–	2.0×10^5	2.1×10^6	–	–
Total	4.0×10^8	1.7×10^9	2.2×10^8	2.3×10^8	8,740	4.7×10^9

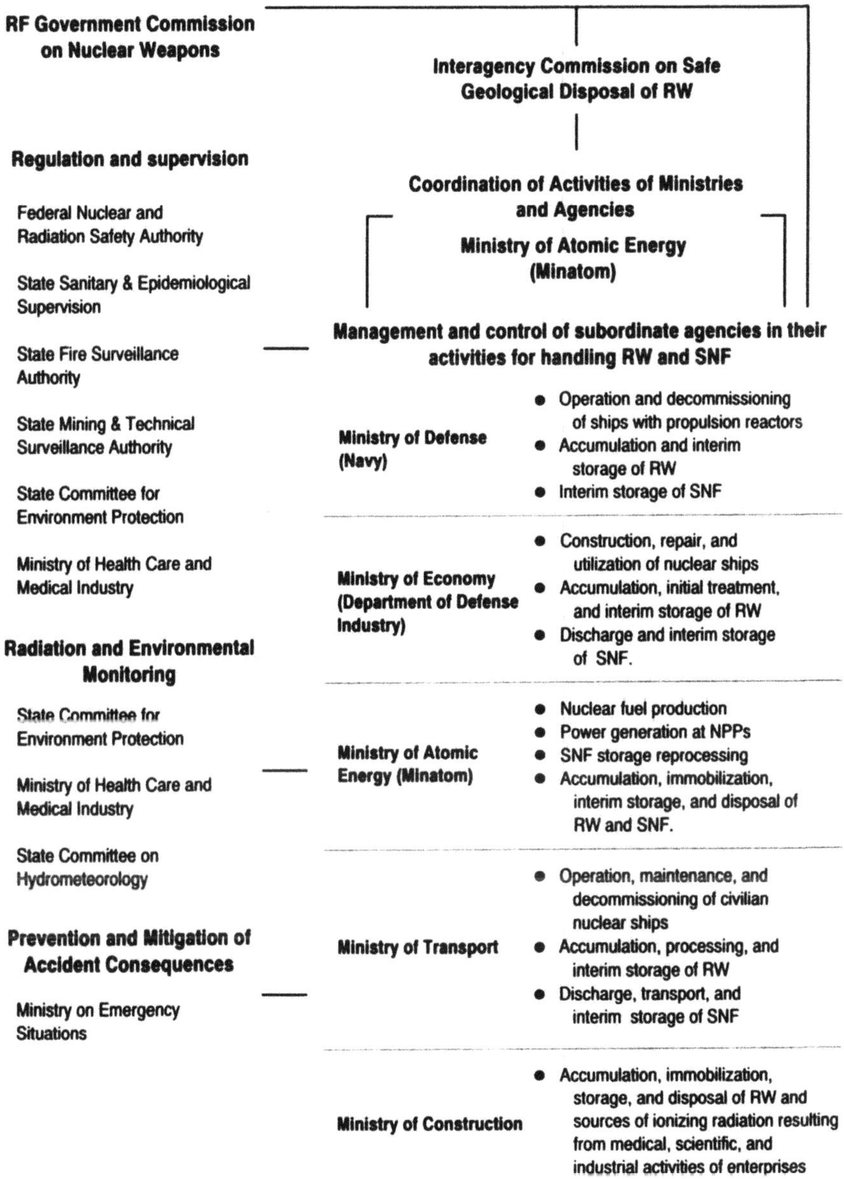

Figure I.1. Structure of state administration of RW management as of 1 January 1996.

Chapter 1

Mining, Processing, and Enrichment of Uranium Ores

1.1 Introduction

At the head-end of the nuclear fuel cycle, the environment is inevitably contaminated by solid, liquid, and gaseous wastes produced in the mining, enrichment, and processing of uranium ores. Radionuclides of three radioactive families (^{238}U, ^{235}U, and ^{232}Th) are released into the atmosphere in the exploitation of uranium deposits. However, the total radioactivity of the released radionuclides results mainly from the ^{238}U family where ^{230}Th, ^{226}Ra, and ^{222}Rn are the major contributors. When compared with the other elements of the nuclear fuel cycle, the exploitation of uranium deposits generates low-level radioactive waste (RW), but it may increase local background radiation for a long period of time (hundreds and even thousands of years).

Solid waste from uranium mines and open pits includes barren rocks (with background or near-background radioactivity), unamenable uranium ores, tailings from radiometric classification of ores, incidentally mined minerals, tailings from heap leaching, etc.

In the case of underground mining, work on capital construction, mine development, and cutting of entries yields 0.2–0.3 or more tons of barren rocks and unamenable ores for each ton of extracted ore. Open pits produce up to 8–10 or more tons of solid waste per ton of ore.

Hydrometallurgical processes (HMPs) turn the mass being processed almost entirely into waste as tailings. Thus, mining and processing of

Table 1.1. Chemical composition of liquid waste from uranium mines.[a]

Component	Concentration (mg/l)	Component	Concentration (mg/l)
Natural uranium	0.3–10.0	Magnesium	10–40
^{226}Ra	1.85–3.7 Bq/l	Iron	0.5–1.0
^{222}Rn[b]	1.85–3.7 Bq/l	Chlorides	2.0–300
Sodium and potassium	40–150	Hydrocarbonates	20–400
Calcium	100–300	Dry residue	500–1,000

[a]Typical pH is about 6.8–7.5.
[b]In equalibrium with ^{226}Ra.

uranium ores produce from 1.3–1.6 tons (mines) to 10–15 tons (open pits) of solid waste per ton of ore.

Liquid waste from uranium mines includes mainly drainable underground water (at the rate of up to 2,000 m^3/day and more), whose composition may vary widely (*Table 1.1*).

Radionuclides (such as uranium, ^{226}Ra, ^{222}Rn, ^{230}Th, ^{210}Po, and ^{210}Pb) accumulated on the ground and in sediments of local water systems as a result of discharge of mine waters are a serious hazard. Their total activity is up to 10–15 Bq/l, whereas maximum permissible concentrations are 0.111 Bq/l.

Tailing dumps are an integral part of the uranium mining industry. The tailings include solid and liquid process wastes. The quantity of the solid waste is roughly equal to the quantity of processed ore, whereas the quantity of liquid waste is two to three times greater. The chemical composition of the liquid fraction in the ore pulp sent to the tailing dumps is dependent on the technology of ore processing and varies widely (*Table 1.2*).

As a source of radioactive contamination, the tailings dump has characteristic features distinguishing it from other industrial sources of contamination. First, the tailings dump is a surface source of uncontrolled release of radioactive aerosols, which has a rather great open surface formed from finely divided dusty materials (sands, sludges). For this reason, the rules for operating the tailings dumps involve measures on reducing the area of dusty surfaces (slopes of embankments, dried surfaces, etc.). Data on the environmental radiation impact of different wastes discharged by the uranium mining industry are given for comparison in *Table 1.3*.

Table 1.2. Chemical composition of the liquid phase of ore pulp from hydrometallurgical plants.[a]

Component	Concentration (mg/l)	Component	Concentration (mg/l)
Natural uranium	0.5–5.0	Manganese oxide	50–100
^{226}Ra	9.9–29.6 Bq/l	Sodium and potassium	200–400
^{210}Po	1.85–55.5 Bq/l	Sulfates	2,000–3,500
Iron dioxide	Trace	Nitrates	500–3,000
Calcium oxide	700–800	Dry residue	8,000–10,000
Magnesium oxide	100–200		
Silicium dioxide	20–40		

[a]Typical pH is about 7.8–8.0.

For the most part, the solid radioactive waste (SRW) represents unamenable ores and mainly HMP tailings. Tailings from acid leaching contain more ^{230}Th than carbonate leaching tailings. The acid and carbonate leaching tailings have almost the same total radioactivity. The specific radioactivity of the HMP tailings is low. Approximately 15% of total radioactivity is removed from the processed ore with the end uranium product. About 70% of initial radioactivity remains in the tailings after the decay of short-lived radionuclides.

Aerosols are dominant in the radioactive contamination of the environment by HMPs. The aerosols fall on all the plant sites and the surrounding areas and contaminate them with long-lived radioactive elements. Concentrated aerosol fallouts of up to 40 Bq/m^2/day may occur at some spots in a plant site.

In addition to traditional mining methods, a method of underground leaching (UL) of uranium was introduced in 1965. UL allows exploitation of poor uranium ore deposits at depths of up to 600 m under adverse geological conditions where the use of traditional mining methods would be economically inefficient. The UL method has been used to mine 9% of prospected uranium resources. When mining by this method, uranium passes into a solution at the site of the the the uranium deposit. The solution is delivered through wells to a sorption plant at the surface. The balance of process solutions is maintained throughout the uranium UL cycle. At the end of the UL process, 0.15–0.18 m^3 of solutions containing radionuclides remain per ton of rock mass in the ore stratum (*Table 1.4*). The bulk of the residual solutions is stored for temporary underground disposal

Table 1.3. Typical parameters of isolation of rehabilitated objects from the environment.

Indices of radiation impact	Dumps of host rocks	Dumps of radiometric classification tailings	Dumps of unamenable ores	Dumps of heap leaching tailings	HMP tailings dumps
Uranium content, %	0.001–0.006	0.010–0.020	0.010–0.030	0.008–0.015	0.001–0.011
Total α activity of rocks, Bq/kg	1.4×10^3–4.4×10^3	3×10^3–1×10^4	3×10^3–1.29×10^4	3×10^3–1×10^4	1.03×10^4–4.9×10^5
γ activity of rocks, μR/hr	20–120	40–250	50–200	50–200	100–1,400
Granulometric composition, mm	Rock mass of run-of-mine size, 400	25–300	Rock mass of run-of-mine size, 300	Rock mass of run-of-mine size, 300	Ground mass, 0.15–0.043
Radon flux density from waste surface, Bq/(m^2 sec)	0.034–0.094	0.08–0.8	0.12–1.2	0.12–1.2	0.2–30
^{226}Ra content of rocks, Bq/kg	37–220	300–550	300–1,700	300–1,700	1,100–27,800
Approximate thickness of rehabilitation layer required to isolate waste from the environment, m	1.0–1.5	1.5–2	1.2–1.6	1.2–1.6	2.5–3.0
Permissible residual release of radon in abandoned objects, Bq/(m^2 sec)	0.2	0.2	0.2	0.2	0.2

Table 1.4. Chemical composition of residual solutions after UL.[a]

Component	Concentration (mg/l)	Component	Concentration (mg/l)
Natural uranium	3–5	Iron	960
^{226}Ra	1.5 Bq/l	Sulfates	12,000–13,000
Sodium and potassium	1,000	Nitrates	50
		Dry residue	20,000
Calcium	480		

[a]Typical pH is about 2.5–3.0.

of low-level liquid radioactive waste (LRW). When the residual solutions move through rocks, the processes of neutralization, precipitation, sorption, ion exchange, and dilution restore solutions to the initial composition of straital waters. The time required to restore the initial composition of the waters depends on the velocity of natural movements of underground waters (15–20 m/yr), and can be determined by calculations and modeling. Soils within the UL site are contaminated with sulfates, nitrates, and natural radionuclides of the uranium series. Thus, current and expected environmental effects of that chemical contamination warrant evaluation.

When the exploitation of a uranium deposit by the UL method is completed, radiation monitoring of the mining site is carried out, and monitoring data for a site rehabilitation project are provided. The contamination of the productive and adjacent horizons is regularly controlled by taking water samples from observation wells. At the final stage of the UL process, the progress in the restoration of the chemical composition of straital waters is tested from observation wells arranged along the flow direction at intervals of 50 m, 100 m, 200 m, etc., within the expected area of propagation of the residual solutions.

Thus, tailing dumps of uranium enterprises contribute the most to environmental contamination at the head-end stages of the nuclear fuel cycle. In this case, ^{222}Rn is the most important contaminant. Within the nuclear fuel cycle as a whole, the contaminating impact of the tailings dumps on the environment is comparable with the impact of nuclear power plants (NPPs), and may have hundreds of times the environmental impact of mining work. Unlike other mining industries, the largest expense in the exploitation of radioactive ore deposits is for reducing, as much as possible, the quantity of radioactive hazardous solid waste from mining works, mine waters, ore dust, and ore enrichment and processing tailings.

1.2 Description of Enterprises

The production complex of the mining and processing enterprises is located in the territories of six Commonwealth of Independent States (CIS) countries (the Russian Federation, the Republic of Ukraine, the Republic of Uzbekistan, the Republic of Kazakhstan, the Republic of Tajikistan, and the Republic of Kyrgyzstan). The complex consists of the following enterprises:

- Argun (Priargunskii) Industrial Mining & Chemical Association (AMCA).
- Lermontov Industrial Association Almaz (IA Almaz).
- Navoi Mining & Metallurgical Combine (Navoi MMC).
- KASKOR Joint Stock Company (until 1992, the Caspian Mining & Metallurgical Combine).
- Industrial Association Tselinnyi Mining & Chemical Combine (IA Tselinnyi MCC).
- Scientific and Industrial Association Eastern Mining and Enrichment Combine (SIA Eastern MEC – based on the Eastern Mining and Enrichment Combine).
- Industrial Association Eastern Combine for Rare-Earth Metals (IA Vostokredmet – until 1992 the Leninabad Mining and Chemical Combine).
- Industrial Association Southern Combine for Polymetals (IA Yuzhpolimetall – based on the Kirgiz Mining Combine).
- Industrial Association Dnieper (Pridneprovskii) Chemical Plant (DCP).

Over time these enterprises have generated a great volume of waste during the mining and processing of uranium ores, such as spoil heaps (188 million tons), HMP tailing dumps, and basins of mine waters (543 million m^3).

In the territory of Russia, the volume of profitable prospected resources amounts to 25% of resources in all of the CIS countries. These natural uranium mining resources are concentrated in the Streltsovsk district, Baikal. In this region, the AMCA conducts uranium mining and processing.

The *Argun Industrial Mining & Chemical Association (AMCA)* is situated in the Krasnokamensk district of the Chita region, 18 km northeast of the town of Krasnokamensk. The AMCA was established in 1968 to mine, enrich, and process uranium-molybdenum ores. Uranium is mined in the Streltsovskii region in three underground mines and in the Tulukui open pit.

Dusty waste sites of amenable and unamenable ores, a uranium open pit, a tailings dump, the Urtui coal pit, an HMP plant for enrichment of uranium and molybdenum, and the Krasnokamensk Heating and Power Plant (HPP) are sources of radioactive contamination for the surrounding territory. The HPP uses Urtui coal, which is high in natural radionuclides (NRNs). During coal burning, NRNs are released into the atmosphere or discharged to an ash disposal site. For this reason the coal's burning is controlled: Only coal with 0.006% or less uranium content is allowed to be used as a fuel, ensuring that permissible levels of NRNs are released, as specified by the NRB–76/87 rules.

Mine waters of the AMCA, industrial sewage from the HPP, and residential sewage are discharged through a collector in surface storage ponds and the local surface hydrologic system. The sewage from the HMP plant is discharged in the tailings dump. The total area of the storage ponds and the tailings dump is 4.264 km^2.

In 1994, the release of uranium radionuclides into Lake Umykei increased by 5.779×10^{10} Bq compared with releases in 1993, but the total radionuclides release was reduced by about 81.4×10^{10} Bq. No excesses over the current standards for releases of radionuclides and their permissible concentrations in the sewage and the collector were recorded.

By using a multistage flow sheet developed at the AMCA for radiometric enrichment of uranium ores, the waste from a radiometric enrichment plant is divided into two classes: waste with sufficiently high uranium to be used for heap leaching and waste with low uranium comparable to the uranium content of rock.

Object	^{210}Pb	^{210}Po	^{226}Ra	^{230}Th	U
Lake Umykei	0.266×10^{10}	0.348×10^{10}	0.189×10^{10}	0.204×10^{10}	8.251×10^{10}
Tailings dump	0.366×10^{10}	0.496×10^{10}	0.259×10^{10}	–	9.731×10^{10}
Total	0.632×10^{10}	0.844×10^{10}	0.448×10^{10}	0.204×10^{10}	17.982×10^{10}

Barren rocks produced in the mining and processing of uranium ores are kept in special dumps. The dumps occupy a considerable area, and

Table 1.5. RW produced at AMCA as of 1 January 1993.

Storage site	Area, 10^3m^2	Volume, 10^3m^3	Mass, 10^3t	Activity, Bq
Dumps of unamenable ores	2,730	104,440	211,260	16.7×10^{14}
HMP tailings dump	4,264	109,000	69,170	29.0×10^{14}
Total	6,994	213,440	280,430	45.7×10^{14}

they are expensive to maintain. The dumps of unamenable ores are used, as a rule, for heap leaching to extract an additional product. Waste from heap leaching is used, in turn, to fill in areas that have already been mined. The dumps of unamenable ores and barren rocks total 2.73 km^2. Before forming a dump, the grading of an appropriate site is made, including topsoil excavation and filling of a 1-m-thick cushion course from barren rock.

The total quantity of RW, including that in the HPP ash disposal site, the coal dump, and a blending stockpile, is 280.4 million tons with a total activity of 45.7×10^{14} Bq (*Table 1.5*).

The tailings dump is located in a natural valley with an embankment and excavated topsoil. The bed of the valley is paved with polyethylene film and a 0.5-m-thick layer of inert soil. The tailings dump is half full.

Within the AMCA site, 18 sources contaminate the atmosphere with ^{222}Rn and 10 sources release natural uranium. In 1994, these sources discharged 2.966×10^{14} Bq of ^{222}Rn and 4.718×10^{10} Bq of uranium into the atmosphere, whereas permissible releases are 41.8×10^{14} and 8.130×10^{10} Bq, respectively. Therefore, no excess of releases above the standards was recorded.

Monitoring and rehabilitation of the areas within the AMCA site, sanitary-protection zone (SPZ), and surveillance zone (SZ) are regularly performed. The total area of contaminated territories in 1994 amounted to 8.483 km^2, including:

Exposure dose rate	Site zone (km^2)	SPZ (km^2)	SZ (km^2)
≤ 60 μR/hr	7.0	0.8	0.418
60 to 120 μR/hr	0.2	0.0136	0.0035
120 to 240 μR/hr	0.0475	–	–

In 1994, radiation monitoring in the AMCA region was carried out by expedition surveys in a 20-km radius of the enterprise as well as with the help of a stationary radiometric network dispersed in a 100-km radius around the AMCA (*Figure 1.1*), and by episodic radioactive fallout monitoring in the vicinity of the Oktyabrskii settlement. The expedition surveys included measurements of exposure dose rates (EDRs) and the sampling of vegetation with subsequent gamma spectrometric analysis. The measured EDRs did not exceed 22 μR/hr at a height of 1 m above the surface. Only cosmogenic ^{7}Be with a specific activity of 110–445 Bq/kg and terragenic ^{40}K with a specific activity of 3,850–10,210 Bq/kg were detected in the vegetation samples. Therefore, the specific activity in vegetation in the neighborhood of the tailings dump near the settlement of Oktyabrskii resulted almost fully from ^{40}K (10,210 Bq/kg). The ^{137}Ce concentrations were below the threshold of detectability.

Observations from the stationary radiometric network show that the average annual EDR ranged from 14 μR/hr to 17 μR/hr (Krasnokamensk).

The average annual total activity of fallout varied from 2.1 Bq/m^{2}d (the village of Dono) to 2.5 Bq/m^{2}d (the settlement of Zabaikalsk). The highest average value, 3.3 Bq/m^{2}d, was observed in Krasnokamensk in July, and the highest average daily value, 25.4 Bq/m^{2}d, was recorded in the same settlement in March.

In 1994, the radioactivity of atmospheric fallout in the vicinity of the AMCA was practically similar to that averaged over the Transbaikal region. As a whole, the AMCA region, like the entire Transbaikal region, is characterized by a steady increase in radioactive contamination of the atmosphere and atmospheric fallouts through airborne dust rich in uranium and thorium series radionuclides .

A special network of hydrogeological observation boreholes was developed at and beyond the site to track possible penetration of mine waters, seepage waters from the surface of the dumps, and leaks through the waterproofing of the tailing dumps into surface and underground waters. These boreholes are used for regular observations. The results obtained to date are satisfactory.

Thus, the present-day uranium-mining enterprise has a radiation im pact on the environment (atmosphere, soil, surface and groundwaters, vegetation) and, as a result, on humans and animals.

As a result, various measures as well as radioecological monitoring are growing in importance. The system must be capable of not only fixing

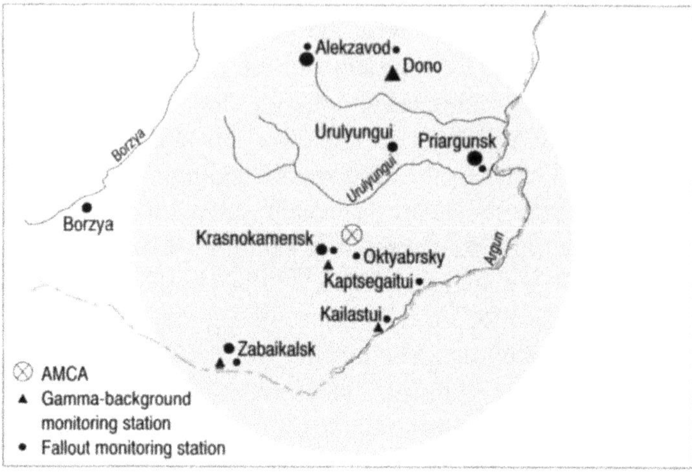

Figure 1.1. Disposition of radiological monitoring stations in a 100-km radius zone around the AMCA.

any disturbances found through monitoring, but also, through a radioecological model, describing the impact of an enterprise on the environment and human health. For this purpose, measures are taken to equip the enterprise's radioecological service with devices and systems for ecological control.

Moreover, a reference network has been developed for complex observations of the soil composition, vegetation, atmosphere, and surface and underground waters. This will ensure prompt detection and elimination of the sources of environmental radiation impact and thereby minimize radiation risk to the population of the region.

Lermontov Industrial Association (IA Almaz) is in the town of Lermontov in the Stavropol Territory. The association was established in 1950 as an industrial uranium-mining enterprise to retrieve deposits at Beshtau and Byk. The association was the leader in the field of geotechnological methods of mining and hydrometallurgy, and it was the first enterprise in the country to use the process of UL of uranium.

The IA Almaz is located in a unique natural zone within the Caucasian mineral waters region. Potential environmental opposition induced the association to search for new directions in the development of the enterprise. Work on complex processing of uranium-bearing phosphorites was initiated in the 1970s. A technology for scandium extraction was developed,

Table 1.6. RW accumulated in IA Almaz as of 1 January 1990.

Storage site	Area, 10^3m^2	Volume, 10^3m^3	Mass, 10^3t	Activity, Bq
Dumps of unamenable ores	570.0	3,960.9	8,403.7	1.40×10^{14}
HMP tailings dump	796.7	12,031.6	14,047.0	1.69×10^{15}
Total	1,366.7	15,992.5	22,450.7	1.83×10^{15}

and pilot production was organized. A shop for ammophos production was built. A shop manufacturing mining equipment was set up as a subdivision of the enterprise and launched production of electrical immersion pumping plants for borehole UL.

Sources of radioactive contamination of the environment were left at the site after completion of uranium ore mining and processing. These waste sites contain unamenable ores and barren rocks as well as a pumped tailings dump. Their total area is 1.367 km^2. The actual volume of unamenable ores and barren rocks in the three dumps is 3,960,900 m^3; the volume of the HMP tailings dump is 12,031,600 m^3 (*Table 1.6*).

A dump of unamenable ores and barren rocks (53,300 m^3 in volume) 4.5 km from the Narta settlement was formed during pilot production and isolated from the environment in trenches about 4 m deep. In mine No. 1 and mine No. 2 (2,107.2 m^3 and 1,800.4 m^3 in volume, respectively), the surfaces of the dumps were rehabilitated (covered with a clay and chernozem layer and reforested). The pumped tailings dump within the HMP site has been operating since 1954 and is now 71% full. Some measures have been taken to diminish the environmental impact of the tailings dump. In particular, the beach of the embankment is isolated by non-RW phosphogypsum produced in the processing of apatite concentrate.

The construction of the *Navoi Mining and Metallurgical Combine (Navoi MMC)* was initiated on the basis of uranium deposits in the Kyzylkum province in 1958. In addition, the construction of the first gold-mining enterprises was also initiated in 1964 on the basis of deposits of auriferous ores. At present, the Navoi MMC comprises mines, open pits, HMP plants, a sulfuric acid production plant, UL site, shops, and other components.

There are three mining departments (MDs) in the structure of the enterprise; the Northern MD, the Southern MD, and the Central MD extract uranium using the UL method.

Table 1.7. RW accumulated in Navoi MMC as of 1 January 1990.

Storage site	Area, 10^3m^2	Volume, 10^3m^3	Mass, 10^3t	Activity, Bq
Dumps of unamenable ores	55.2	n.a.	166	3.4×10^{12}
HMP tailings dump	6,022.0	47,800.0	52,800	7.4×10^{15}
UL sites	3,831.0	9,126.4	n.a.	2×10^{12}

n.a. = not available.

In the Northern MD, the UL sites that had occupied an area of 1,283 m^2, 1.0–1.5 km from the town of Uchkuduk, were decommissioned between 1975 and 1989. After the implementation of environmental protection measures, the surface EDR within the SPZ (3.52 km^2) has not exceeded 20 μR/hr. The volume of residual solutions in the depths is no more than 2,392,000 m^3.

In the Southern MD, five UL sites, 1.0–60 km from the settlement of Nurabad, with a total area of 2,548 m^2, were decommissioned between 1982 and 1990. Since the sites were rehabilitated, the EDR within the SPZ (3.539 km^2) has not exceeded 19–25 μR/hr. The volume of residual solutions left in the depths is 4,389,800 m^3.

Onsite dumps of unamenable ores were built during the operation period of a mine 1.5 km from the Nurabad settlement. The mine was in operation from 1965 to 1982. The volume of rock and ore mass amounted to 166,700 tons. The EDR within the SPZ around the dumps is 17 μR/hr.

In the site of the Central MD, 15 km from the town of Zarafshan, uranium was mined between 1979 and 1990. Since the site was rehabilitated, the EDR within the SPZ has not exceeded 20 μR/hr. The volume of residual process solutions in the productive underground horizon has totaled 2,344,600 m^3 at an activity of 1.0×10^{12} Bq.

An HMP flood tailings dump of over 6.022 km^2, located 1 km from the settlement of Durmyan, has been in operation since 1964. This tailings dump has a recycling water supply system and hydraulic feed of the HMP tailings when the fluid fraction of the pulp returns into the HMP plant. To date the tailings dump has accumulated 52,800,000 tons of waste containing uranium, radium, and polonium. The EDR in the site has not exceeded 30 μR/hr. The SPZ area is 8.360 km^2.

The *KASKOR Joint Stock Company*, which was created in 1992 from the Caspian Mining and Metallurgical Combine (MMC), is a large-scale ore-mining and ore-processing enterprise that comprises 28 divisions.

Table 1.8. RW accumulated in KASKOR Joint Stock Company as of 1 January 1990.

Storage site[a]	Area, $10^3 m^2$	Volume, $10^3 m^3$	Mass, $10^3 t$	Exposure dose rate, μR/hr	Activity, Bq
Burial ground	100	1.8	6	40–60	–
CHMP tailings dump	75,000	210,000	68,100	15–20	6.3×10^{13}
GREC storage facility	90	904	–	250	–

[a]Each of the storage sites is surrounded by a 0.5-km radius SPZ.

The Caspian MMC was established in 1959 on the basis of uranium-phosphorus deposits in the inhabited territory of the Mangyshlak Peninsula in the Republic of Kazakhstan.

One of the KASKOR divisions, a major chemico-hydrometallurgical processes (CHMP) plant, is a large enterprise performing complex processing of local uranium-phosphorus ores and graphite concentrate from the Kola Peninsula. Its basic products are uranium concentrate, scandium oxide and fluoride, crystalline scandium, and alumino-cadmium hardeners as well as rare earth concentrates of light- and medium-weight groups. The plant produces fodder dicalcium phosphate (precipitate), food dicalcium phosphate (dentaphos), and ammonium sulfate from large tonnage process waste.

During its operation, the enterprise accumulated 210,904,000 m^3 of RW in a CHMP tailings dump, a burial ground, and a gypsum and rare earth concentrate (GREC) storage facility (*Table 1.8*).

The CHMP tailings dump is located 5 km north of the plant site in a natural hollow. During the 27-year operation of the tailings dump, the volume of waste in it totaled 210,000 m^3 – that is, 70% of the design capacity of the dump. The CHMP pulp directed to the tailings dump contains radionuclides ^{226}Ra and ^{210}Bi. The radiometric survey of soils and the upper aeration layer 0.3 m beneath the surface revealed no contamination of radionuclides in the territory surrounding the tailings dump. An excess above the natural background is observed along the shoreline of the tailings dump where the EDR ranges from 14–60 μR/hr, with the maximum value corresponding to the uncovered tailings sediments.

The burial ground is 12 km from the town of Shevchenko and 6 km from the CHMP. The burial ground was in service for 25 years before it was filled. It accommodates 6,000 tons of RW containing radionuclides ^{210}Po, ^{230}Th, ^{238}U, and others. The surface and bottom of the burial

Table 1.9. RW accumulated in IA Tselinnyi MCC as of 1 January 1990.

Storage site	Area, 10^3m^2	Volume, 10^3m^3	Mass, 10^3t	Activity, Bq
Dumps of unamenable ores	823	14,592	16,200	4.5×10^{14}
HMP tailings dump	7,220	52,000	56,600	2.0×10^{15}
Total	8,043	66,592	72,800	2.4×10^{15}

ground are isolated from the environment by a compacted clay layer. The EDR at the surface of the burial ground and within its SPZ does not exceed 40–60 μR/hr.

The GREC storage facility, 10 km from the town, is waterproofed by isobutylene and concrete. It has a design capacity of 29,626 m^3 and holds 904 m^3 of SRW containing ^{210}Pb, ^{227}Ac, and ^{228}Ac. The EDR in the vicinity of the storage facility is 250 μR/hr.

The *Industrial Association Tselinnyi Mining & Chemical Combine (IA Tselinnyi MCC)* was established on the basis of large uranium, uranium-molybdenum, and uranium-phosphorus ore deposits in North Kazakhstan. Its construction was started in 1956. The basic production facilities are located 20 km from the town of Stepnogorsk. The uranium-mining enterprises are located 120–500 km from the basic production facilities.

IA Tselinnyi MCC is comprised of an HMP plant, a chemical plant, a heap leaching facility, an equipment maintenance and repair shop, a geologic field crew, and power and heating plants.

During its operation, the facilities of the association produced 66,592,000 m^3 of RW that accumulated in a flooding HMP tailings dump and onsite dumps of unamenable ores in the mining enterprises (*Table 1.9*).

The flooding tailings dump occupies 7,220 m^2 and is located in the SPZ of the HMP plant. It was used between 1969 and 1990 and filled to 90% of its capacity. The bottom of the tailings dump is paved with polyethylene film. The tailings dump consists of two ponds. The surface of the one pond is isolated from the environment by polyethylene film. The second, an evaporating pond, has an open surface. The basic radioactive contaminants in the HMP pulp are the natural radionuclides of the ^{238}U and ^{232}Th series.

SRW from radiometric enrichment and heap leaching is stored in piles within the SPZ, 0.5–7.0 km from the enterprise. The piles are placed on

Table 1.10. RW accumulated in SIA Eastern MEC as of 1 January 1990.

Storage site	Area, $10^3 m^2$	Volume, $10^3 m^3$	Mass, $10^3 t$	Activity, Bq
Dumps of unamenable ores	181.4	1,960	3,770	3.4×10^{13}
HMP tailings dumps	2,859.0	37,600	37,750	2.1×10^{15}
Total	3,041.4	39,560	41,520	2.1×10^{15}
UL sites	2,250.0	11,235	n.a.	$(1.8–1.9)\times10^{11}$

n.a. = not available.

a surface that has been compacted by rolling, without implementation of special waterproofing measures. The EDR on the surface of the piles does not exceed 100 μR/hr. The bottom of the 628,000-m^3 heap-leaching dump is covered by asphalt and film to facilitate the collection of productive solutions in sprinkled ore mass and to prevent contamination of the soil.

The *Scientific and Industrial Association Eastern Mining and Enrichment Combine (SIA Eastern MEC)* succeeded the Eastern Mining and Enrichment Combine founded in 1951 on the basis of the Pervomaiskii and Zheltorechenskii uranium-containing iron ore deposits in the Kirovograd ore region in the Ukraine. Uranium and iron were extracted from these mines. As of 1990, the enterprise stored 3,770,000 tons of unamenable ores and barren rocks in dumps in an area of 0.181 km^2 (*Table 1.10*).

The bottom of the dumps is formed from loams. Some 1,960,000 m^3 of unamenable ores were buried in the cavities of the Novaya iron ore mine. This amounts to 8% of the enterprise's capacity.

Since 1976, a tailings dump with a useful volume of 29,200,000 m^3 has been in service 7 km from the HMP plant. As of 1990, it was 87% full, corresponding to 25,404,000 m^3 of waste in the form of pulp. The bottom of the tailings dump is waterproofed with compacted natural loam. The EDR in a 950-m-radius supervised zone varies in the range of 17–25 μR/hr. Another tailings dump 400 m from the HMP plant had been in service since 1966. At present it has been taken out of service because it was completely filled (8,400,000 m^3). The tailings dump was constructed with no waterproofing course. The EDR within the supervised zone is no more than 25 μR/hr.

SIA Eastern MEC mined uranium using the UL method from the Devladovo and Bratskoe deposits in sedimentary rocks of the Buchak productive horizon enclosed between confining clay horizons.

A UL site on the Devladovo deposit is 5 km from the Devladovo station of the Pridneprovskaya Railway. The enterprise was in service from 1962 to 1983. The volume of residual process solutions in the productive horizon totaled 6,035,000 m^3 and has an activity of 1.8×10^{11} Bq. The predicted self-purification time for the water-bearing horizon is 280 years at a natural underground water velocity of 20 m/yr. The 15-m-thick productive horizon is 60–65 m deep.

A UL site on the Bratskoe deposit is in the Nikolaev region 10 km from the Bratskoe settlement. The deposit was in operation between 1971 and 1989. The volume of residual process solutions in the productive horizon amounted to 5,200,000 m^3 and has an activity of 1.9×10^{11} Bq. The predicted self-purification time for the water-bearing horizon is 280 years at a natural underground water velocity of 20 m/yr. The 15-m-thick productive horizon is at a depth of 94–100 m.

The surfaces of the two UL sites were remediated. The remediation works included removal of 142,600 m^3 of soil with an activity of $(1.6–1.8) \times 10^{11}$ Bq and its subsequent burial in trenches 2.8–3.5 m deep.

At present SIA Eastern MEC comprises mining enterprises, an HMP plant, a plant for production of sulfuric acid, and some service enterprises.

The top-priority activities of SIA Eastern MEC are mining of uranium and iron ores as well as production of iron ore and uranium concentrates and sulfuric acid.

The *Industrial Association Eastern Combine for Rare-Earth Metals (IA Vostokredmet)* was established on the basis of the Leninabad Mining and Chemical Combine, and is situated in the western part of the Fergana Valley, one of the most picturesque areas of Central Asia.

IA Vostokredmet includes an HMP plant, an engineering plant, research laboratories of chemical processes and automated management systems, an instrumentation and automatic equipment laboratory, a geological maintenance and repair shop, and a construction and mounting department. IA Vostokredmet is one of the leading enterprises in the field of UL and heap leaching of metals.

The combine was established in 1945 as a large-scale uranium-mining enterprise on the basis of uranium deposits in Tajikistan, Kyrgyzstan, and Uzbekistan. As the country's first raw materials base, IA Vostokredmet incorporated seven mines and five plants, including a hydrometallurgical shop at Plant B, and the Tabashar, Adrasman, Mailisai, Uigur, and Tyuya-Muyun mines.

Table 1.11. RW accumulated in IA Vostokredmet as of 1 January 1990.

Storage site	Area, 10^3m^2	Volume, 10^3m^3	Mass, 10^3t	Activity, Bq
Dumps of unamenable ores	224	2,678	1,847	7.0×10^{12}
HMP tailings dumps	1,741	32,229	33,684	2.5×10^{14}
UL sites	n.a.	14,754	n.a.	4.0×10^{12}

n.a. = not available.

The settlement of Chkalovsk was built 10 km from the city of Khodzhent (Leninabad). Later, IA Vostokredmet's basic HMP was built in Chkalovsk to process uranium ores from the different deposits. A pilot hydrometallurgical shop in Tabashary (40 km from Chkalovsk) was converted to a pilot HMP plant where ores from the Meilisui, Uigur, and Adrasman mines and deposits from other regions have been processed in addition to the Tabashary ores.

Throughout the period of IA Vostokredmet's activities, 34,907,000 m^3 of low-level RW accumulated in its territory. The wastes are stored in nine tailings dumps (a total area of 1.741 km^2) and 21 dumps of unamenable ores (a total area of 0.224 km^2) in the mining and processing enterprises (*Table 1.11*).

In two mining departments, which extracted uranium by the UL method, the volume of residual process solutions in the productive horizons at a depth of 450–600 m amounted to 14,754,000 m^3 and has an activity of 4.0×10^{12} Bq.

The dumps of unamenable ores and barren rocks are dispersed over the vast territory of IA Vostokredmet. Twelve dumps of MD II are arranged on the terraces of the Chauli-Sai River, 0.3–6.5 km from the town of Khodzhent. The quantity of RW accumulated within the period of operation of the dumps (1953–1980) is 469,800 tons. The dumps have an area of 0.157 km^2 and an activity of 4×10^{12} Bq. The EDR within the SPZ is 200–600 μR/hr.

Eight dumps, 1–10 km from the settlement of Charkassar, were formed within the operation period (1954–1989) of the Adrasman mine. Their total area is 34,000 m^2; the RW totals 292,100 tons and has a total activity of 5.4×10^{11} Bq. The EDR within the SPZ varies from 70–100 μR/hr. On the site of the mine at a distance of 50 m from the settlement there is a tailings dump of an ore-mining enterprise. The tailings dump, which occupies an area of 34,000 m^2, was in operation between 1949 and

1958. The volume of waste in it is 102,000 m^3, and it has a total activity of 2.4×10^{12} Bq. The EDR within the SPZ is 50–60 μR/hr.

A storage of poor ores with a volume of 1,195,000 m^3 and an activity of 2.5×10^{12} Bq remained in the site of the Tabashar mine, 4 km from the settlement Tabashar after the 15-year (1950–1965) operation of the mine. The EDR within the SPZ is 40–100 μR/hr. Four tailings dumps, with a total volume of 4,547,000 m^3 and a total area of 0.574 km^2, were formed at different times during the operation of a pilot HMP plant (1949–1965). The total activity of waste in the tailings dumps is 3.6×10^{13} Bq. The EDR within the SPZ is 40–60 μR/hr.

In the valley of the Maili-Su River, 15 km from the town of Mailisai, an HMP tailings dump was in operation from 1947 to 1962. At present, the surface of the tailings dump is isolated from the environment by a 0.5-m-thick soil layer. The volume of waste in the tailings dump is 1,340,000 m^3; the activity is 1.9×10^{13} Bq. The EDR on the surface of the tailings dump is up to 30 μR/hr.

A tailings dump for the waste from HMP Plant I has been in operation since 1963 in the Digmai cavity, 4 km from the town of Gafurov. The tailings dump occupies 0.692 km^2 and is filled with 19,200,000 m^3 of waste. The activity of the waste amounts to 1.56×10^{14} Bq. The EDR within a 1-km-radius SPZ varies from 40–250 μR/hr.

From 1949 to 1967, an HMP tailings dump, with an area of 0.267 km^2, 2 km from the town of Gafurov, was in operation. The volume of waste is 2,600,000 m^3; the activity of waste is 2.9×10^{13} Bq. A 0.5-m-thick soil layer covers the surface of the tailings dump. The EDR in the vicinity of the tailings dump varies from 20–60 μR/hr.

A tailings dump in the town of Gafurov operated between 1945 and 1950. It was intended as a pilot plant for hydrometallurgical processing of uranium ores. The tailings dump (an area of 39,000 m^2) is disposed within the area of service and pilot production buildings. It was completely dried, and in 1963 its surface was covered with a 1.0-m-thick layer of inert soil. The radon concentration in the air directly above the dump was between 2–5 Bq/m^3; the release of radon to the atmosphere from the entire surface of the dump is about 2.96×10^{12} Bq/yr. A decision about the onsite burial of the tailings dump was made in 1991. The most effective measures were chosen to ensure the reliable disposal of the tailings dump and the radiation safety of the surrounding areas. The choice of an appropriate material for covering the dump surface is under study. The covering must

Table 1.12. RW accumulated in IA Yuzhpolimetall as of 1 January 1990.

Storage site	Area, 10^3m^2	Volume, 10^3m^3	Mass, 10^3t	Activity, Bq
Dumps of unamenable ores	2,144	55,436	110,873	6.0×10^{14}
HMP tailings dumps	2,576	35,695	34,461	3.2×10^{15}
Total	4,720	91,131	145,334	3.8×10^{15}
UL sites	424	2,009	n.a.	3.3×10^{11}

n.a. = not available.

reduce the release of radon from the surface down to 0.05–0.1 Bq/(m^{2}sec). Loessial loams with subsequent compacting have been suggested for this purpose. The thickness of the loam layer must be 1.0–2.5 m to ensure that the time of radon diffusion to the surface would be at least 30 days until the almost complete decay of the radon.

The *Industrial Association Southern Combine for Polymetals (IA Yuzhpolimetall)* was established on the basis of the Kirgiz Mining Combine founded in the 1960s. The Combine mined and processed uranium ores from the southern Kazakhstan ore region.

While IA Yuzhpolimetall was operating, 18 dumps of unamenable ores and barren rocks with a total volume of 55,436,000 m^3 and four tailings dumps containing a total of 35,695,000 m^3 of radioactive pulp were formed on its sites (*Table 1.12*). Some 2,009,000 m^3 of residual process solutions have been left in the depths of two depleted UL sites. The solutions contain the natural radionuclides of the uranium series and high concentrations of chemical components. The dumps of unamenable ores and rocks are 80 km from populated areas. All of the dumps had been taken out of operation by 1991. No data on the rehabilitation of the spaces occupied by the dumps are available. Three tailings dumps between 2–11 km from the settlement of Minkush were taken out of service between 1960 and 1969. A 0.5-m-thick layer of loam was filled and compacted on the surfaces of the tailings dumps as an environmental protection measure. The EDR within the SPZ is no more than 30 μR/hr.

An onsite tailings dump of the Kara-Baltinskii Mining Combine (an independent industrial enterprise since 1992) has been in service since 1955. Some 65% of the beach embankment is formed from loam, and 35% is covered with polyethylene film. As of 1995, the tailings dump was filled to 54% of the design capacity. At present, the tailings dump is used

Table 1.13. RW accumulated in IA DCP as of 1 January 1990.

Storage site	Area, 10^3m^2	Volume, 10^3m^3	Mass, 10^3t	Activity, Bq
HMP tailings dumps	3,998	13,663	52,017	2.7×10^{15}

in reprocessing molybdenum-tungsten-bearing raw materials and RW in an HMP plant.

The *Industrial Association Dnieper Chemical Plant* (DCP) started to process uranium-containing raw materials from iron ore deposits of the Krivorozhskii field in 1948. From 1948 to 1954, uncontrolled tailings dumps were created in natural clay pits within the SPZ of the enterprise.

Throughout DCP's activities, 13,663,000 m^3 of HMP tailings accumulated in the tailings dumps (*Table 1.13*). The tailings dumps have no protective barriers. Another tailings dump (D) of the plain-flooding type, with a capacity of 10 million m^3 and an area of 76 ha, was put into service in 1954 in the Dnieper flood lands to store RW from the hydrometallurgical production of uranium. The tailings dump has no waterproofing means. Its embankment is made of waste from a coke plant by-product – namely, crushed stone, gravel, and silty soils from coal, clay, and mica shales. After compaction of its beaches in 1970, the tailings dump was filled with phosphogypsum from processing of apatite.

The first section of a flooding-type tailings dump (C), with a capacity of 8.7 million m^3 and 14 km from the city of Dneprodzerzhinsk, was put into service in 1968. An asphalt-bitumen impervious blanket is used as an antiseepage measure. The tailings dump operated with a circulating water supply and did not discharge wastewater into the hydrosphere. The second section, with a capacity of 19.2 million m^3 and antiseepage protection, was put into service in 1983 after the first section was completely filled. The reliability of the antiseepage measures was confirmed by full-scale studies and regime observations.

A radioactive tailings dump (DP-6) is stored in a 54×121-m trench, which is covered with solid rubble and a 1.5–2.0-m-thick chernozem layer, as well as sowed with grass. (The trench originally stored the remains of a blast furnace used from 1948–1964 to melt uranium ores.) The archival materials of the enterprise, which are not complete, were used to calculate the total activity of the tailings dumps. The survey work is now under way to determine reliable data on the activity of the tailings dumps.

Data on two uranium mining and processing enterprises in the Russian Federation were included in a simple Radleg operational database. The prototype for development of this database was made from a departmental database on RW storage and disposal facilities in nuclear fuel cycle (NFC) enterprises, which is based on RW inventories carried out in 1990 and 1993.

Information on the enterprises located in the territories of the CIS – the Republic of Ukraine, the Republic of Uzbekistan, the Republic of Kazakhstan, the Republic of Tajikistan, and the Republic Kyrgyzstan – will be included in the database during the second phase of the project.

1.3 Conclusions

As a result of many years of uranium ore mining and processing, large volumes of low-level RW, containing natural radionuclides of ^{238}U and ^{232}Th families, have been generated in the form of rock dumps (138 million m^3), tailings dumps of hydrometallurgical plants, and mine water pools (543 million m^3). The total area of these sites is 130 km^2.

Table 1.14 shows the amount of materials and the total radioactivity at different sites. The wastes of both mining and processing of the ores contain long-lived radionuclides: ^{238}U ($T_{1/2} = 4.47 \times 10^9$ years), ^{230}Th ($T_{1/2} = 7.7 \times 10^4$ years), and ^{226}Ra ($T_{1/2} = 1.6 \times 10^3$ years), transforming continuously to ^{222}Rn ($T_{1/2} = 3.8$ days). For this reason, due to presence of the above-mentioned radionuclides, the radiation background in underground uranium mines, quarries, industrial sites of ore processing enterprises, and adjacent areas, exceeds the natural background by several times.

The main source of radioactive contamination – ore wastes – are generated as a result of ore processing and are collected in tailings dumps. Within the nuclear fuel cycle, the contaminating impact of the tailings dumps on the environment is comparable with the impact of NPPs and exceeds from tens to hundreds of times the environmental impact of mining work per se. As a source of radioactive contamination, the tailings dumps have characteristic features that distinguish them from other industrial sources of contamination. First, a tailings dump is a surface formed from finely divided dusty materials (sands and sludges). For this reason, the rules for operating the tailings dumps involve measures on reducing the area of dusty surfaces (slopes of embankments, dried surfaces, etc.).

Table 1.14. Summary data on RW mass and activity accumulated at uranium ore mining and processing industry enterprises of the FSU, as of 1 January 1990.

	Dumps of unamenable ores		Tailings dumps	
Enterprise	RW mass, 10^3 t	RW activity, 10^{14} Bq	RW mass, 10^3 t	RW activity, 10^{14} Bq
AMCA[a]	211,260	16.7	69,170	29.0
IA Almaz	8,403	1.4	14,047	16.9
Navoi MMC	166	0.034	52,800	74.0
KASKOR JSC	n.a.	n.a.	68,145	0.629
IA Tselinnyi MCC	16,200	4.5	56,600	20.0
IA Vostokredmet	1,847	0.07	33,684	2.5
SIA Eastern MEC	3,770	0.34	37,750	21.0
IA Yuzhpolimetall	110,873	6.0	34,461	32.0
IA DCP	n.a.	n.a.	52,017	27.0
Total	352,519	29.0	418,674	223.0

[a] As of 1 January 1993.
n.a. = not available.

Data on the environmental radiation impact of different wastes discharged by the uranium mining industry are given in *Table 1.3*.

To meet standards for the residual radiation level (0.2 Bq m^2sec), the approximate depth of remediation layer for reliable isolation of the RW from the environment must be not less than 1.0–2.0 m for dumps of barren rocks and 2.5–3.0 m for tailings dumps. Storage of uranium ore mining and processing RW require continuous monitoring, which is conducted within the limits of radiation control, SPZs, and SZs of these objects.

A federal program entitled "Management of Radioactive Waste and Spent Nuclear Materials, Their Utilization and Disposal for 1996–2005" (hereafter, "the Program") was approved by the resolution of the Russian Federation in 1995 to comprehensively solve RW management problems as well as to construct long-term storage or disposal facilities for reliable isolation of RW from the biosphere. For uranium mining and processing facilities, the Program envisages:

- A comprehensive study of the environmental impact of uranium mining facilities and the shut down or, if necessary, disposal of decommissioned mines, open pits, and tailings dumps.

- Development of plants for decontamination and processing of LRW and SRW produced in uranium ore mining and processing.
- Development of plants for reprocessing of waste from enriched uranium production to extract useful elements and reduce the volume of the waste.

Even if the uranium mining and processing enterprises are abandoned or put in stand-by mode, enterprise personnel and local populations must still be protected from radiation hazards. Among the factors of significance are releases of mine waters and the liquid phase of the HMP ore pulp to the hydrosphere; releases of radon and products of its decay into the atmosphere; and dusts from dumps of unamenable ores, enrichment tailings, and host rocks. For this purpose, an initiative called "Sanitary Regulations on Abandonment, Laying-up, and Conversion of Radioactive Ore Mining and Processing Enterprises" has been worked out and approved. This document establishes the permissible residual levels of radiation contamination of the facilities that will be abandoned or shut down (*Table 1.15*). The standards worked out for the residual radiation contamination ensure that the radiation risk to the facilities' personnel and the population working or living in the neighboring regions will be limited. In addition to reduction in the activity of the corresponding radiation source, reliable isolation of RW, especially its solid phase, from the environment is of extreme importance. For this purpose, dumps of barren rocks, unamenable ores, radiometric enrichment tailings, and heap leaching tailings, as well as HMP tailings dumps, must be reliably covered with a remediation layer of inert rocks (*Table 1.3*). In so doing, the residual radon release must not exceed 1 Bq/(m^2sec) for standby enterprises and 0.2 Bq/(m^2sec) for abandoned enterprises.

To ensure the radiation safety during the abandonment process, standby mode, or conversion of uranium mining and processing enterprises, the following measures must be implemented:

- A radiation–ecological survey of the actual state of the enterprises.
- Research and development work involving the assessment of the environmental impact of the enterprises.
- Development of abandonment (standby) projects including both an assessment of environmental damage and a feasibility study of planned environmental protection measures.

Table 1.15. Residual radioactive contamination of some objects in uranium mining and processing enterprises to be abandoned.

Objects of enterprises to be abandoned, put on standby, or converted	Permissible radioactive contamination		
	γ-radiation dose rate, μR/hr	Radon concentration, Bq/m^3	Specific α activity, Bq/kg
Buildings, surface complexes, structures, waters, process equipment	<70	<200	n.a.
Scrap, equipment to be reused	<70	n.a.	n.a.
Office buildings, dining rooms, workshops, etc. Shops to be used for other purposes	<35	<200	n.a.
Lands to be rehabilitated after removal of ore storage facilities, onsite leaching complexes, spilled ores, and chemicals:			
For farming	<20 above background	0.2 Bq/(m^{2}sec)	<600 at a depth of 100 cm; up to 7,400 at some spots
For forestry	<20 above background	0.4 Bq/(m^{2}sec)	<1,200 at a depth of 100 cm; up to 7,400 at some spots
For industrial and civil construction	<20 above background	0.1 Bq/(m^{2}sec)	<1,200 at a depth of 200 cm; up to 7,400 at some spots
For sanitary–hygienic purposes	<20 above background at 100-cm height	0.4 Bq/(m^{2}sec)	<1,200 at a depth of 200 cm; up to 7,400 at some spots
Dumps of barren rocks, unamenable ores, radiometric classification, and heap leaching tailings	<80 above background at 100-cm height	1.0 Bq/(m^{2}sec)	n.a.
HMP tailings dumps	<100 above background	1.0 Bq/(m^{2}sec)	n.a.
Mine waters	n.a.	111	n.a.

n.a. = not available.

The concepts, methods, and normative criteria of radiation safety formulated in the Program give a guarantee for the timely and reliable execution of abandonment, standby, and conversion work, as well as the recommissioning of the enterprises.

Bibliography

Iskra, A.F., and Lutsenko, I.K., 1991, Ecological, technological, and economic criteria for use of water resources and protection of surface and underground waters from exposure to radioactive and toxic substances, In: *The 2nd Conference of the Nuclear Society on Radioactive Waste: Problems and Solutions: Part I*, pp. 181–189, Moscow, Russia.

ISTC (International Science and Technology Center), 1996a, Phase I: The Creation of a Simple Operational Database with Connection to GIS, Including Presently Available Data on the Radiation Legacy of the ex-USSR, In: *The Final Technical Report on the ISTC Project #245 Radleg*, ISTC, Moscow, Russia.

ISTC (International Science and Technology Center), 1996b, Radleg simple operational database (tables 1, 2, 3, 4, 6, 10, 11, and 12), ISTC, Moscow, Russia.

Kedrovskii, O.L., and Skorovarov, D.I., 1987, *Mining and Processing of Uranium Ores. Nuclear Science and Engineering in the USSR*. Energoatomizdat, Moscow, Russia.

Kruglov, A.K., 1995, *Development of Nuclear Industry in the USSR*, TsNIIAtominform, Moscow, Russia.

Makhonko, K.P., ed., 1995, *Radiation Situation in the Territory of Russia and Neighboring States in 1994: A Yearbook*, NPO Taifun, Obninsk, Russia.

Minatom (Ministry of the Russian Federation for Atomic Energy), 1994, *An Interdepartmental Instruction on Protection of the Earth Surface and Rehabilitation of Areas Damaged in Mining and Processing of Radioactive Ores* (MI ORZ–94), Minatom, Moscow, Russia.

Minatom (Ministry of the Russian Federation for Atomic Energy), 1993, *The Atomredmetzoloto State Concern* (an advertising booklet), The Central Research Institute of Management, Economy, and Information of Minatom, Moscow, Russia.

Mosinets, V.N., 1991a, Radioactive waste of uranium mining plants and its ecological impact, In: *The 2nd Conference of the Nuclear Society on Radioactive Waste: Problems and Solutions: Part I*, pp. 79–87, Moscow, Russia.

Mosinets, V.N., 1991b, Radiation safety in abandonment, laying-up, and conversion of uranium mining enterprises, In: *The 2nd Conference of the*

Nuclear Society on Radioactive Waste: Problems and Solutions: Part I, pp. 116–121, Moscow, Russia.

Russian Federation Special Program, 1995, *Management of Radioactive Waste and Spent Nuclear Materials, Their Utilization and Disposal for 1996–2005* (approved by a decision of the Russian Federation, No. 1030), Russian Federation Special Program, Moscow, Russia.

USSR Ministry of Health, 1991, *Sanitary Regulations on Abandonment, Laying-up, and Conversion of Radioactive Ore Mining and Processing Enterprises* (SPLKP–91), No. 6028–91, Moscow, Russia.

VNIICHT Files, 1991, *Documentation on Departmental Inventory of Radioactive Waste and Its Storage and Disposal Sites*, 1 January 1990, VNIICHT, Moscow, Russia.

VNIICHT Files, 1994, *Documentation on Departmental Inventory of Radioactive Waste and Its Storage and Disposal Sites*, 1 January 1993, VNIICHT, Moscow, Russia.

Zaitsev, V.F., *et al.*, 1991, Radioecological problems of mining and processing of uranium ores, In: *The 2nd Conference of the Nuclear Society on Radioactive Waste: Problems and Solutions: Part I*, pp. 138–142, Moscow, Russia.

Chapter 2

The Industrial Complex for Uranium Enrichment

2.1 Introduction

Initially, the industrial complex for uranium enrichment in the Soviet Union was created for making nuclear weapons. Later, as civilian nuclear power increased, the capacity of the complex was partly reconstructed for production of low-enriched uranium (LEU) to supply nuclear power plants (NPPs) with reactor fuel. In 1988, the production of highly enriched uranium (HEU) in the Soviet Union ceased. At present, the complex for uranium enrichment is used only for nuclear fuel production. LEU is delivered now both to the Russian nuclear fuel market and abroad.

The industrial complex for uranium enrichment includes uranium hexafluoride (UF_6) production plants and uranium enrichment plants.

The plants have special departments for utilization, processing, and storage of uranium radioactive wastes (RWs) generated during operations. All the plants work in one balanced cycle; that is, utilization of the plants for both primary UF_6 and LEU production is regulated by demands of the internal market of Russia and obligations of international contracts.

On 1 January 1997, two plants for primary UF_6 production and four plants for uranium enrichment were operating in four locations in Russia:

- The Urals Electrochemical Combine (UECC) in Novouralsk, Sverdlovsk region, has operated an enrichment plant since 1949.

- The Electrochemical Plant (ECP) in Zelenogorsk, Krasnoyarsk land, has operated an enrichment plant since 1962.
- The Siberian Chemical Combine (SCC) in Seversk, Tomsk region, has operated an enrichment plant since 1953 and a UF_6 production plant since 1954.
- The Angarsk Electrolysis Chemical Combine (AECC) in Angarsk, Irkutsk region, has operated an enrichment plant since 1957 and a UF_6 production plant since 1960.

All the enterprises listed are the state's property and are administered by the Ministry of the Russian Federation for Atomic Energy (Minatom). They have permission to conduct independent commercial business. No UF_6 production and uranium isotope enrichment enterprises exist in other countries of the former Soviet Union (FSU).

Along with fulfilling their main task – primary and low-enriched UF_6 production – the enterprises are also engaged in related activities: storage and transportation (in packed form) of nuclear materials (uranium), process equipment decontamination, uranium RW management including operation of uranium waste storage and disposal facilities on the enterprises' territory, etc.

The production capacities of the industrial complex for uranium enrichment are distributed as follows:

UF_6 production plants	
AECC	67%
SCC	33%
Uranium enrichment plants	
UECC	49%
ECP	29%
SCC	14%
AECC	8%

All the enrichment plants started with gas-diffusion technology. In 1962 they began intensively using gas centrifuges. In 1992, industrial operation of the gas-diffusion technology was completely stopped, and now only gas centrifuges are in use at the enrichment plants. The gas centrifuges do not need repair for their entire service life, which is not less than 15 years. Correspondingly, opening of the internal process space is not necessary.

The uranium enrichment process technology is continuous and highly automated. The process is carried out in large gas-tight systems under extreme vacuum conditions. The potential impact of the enrichment plant on the environment is at the same level as that of uranium and thorium mining enterprises.

The UF_6 production plants include three main subdivisions:

- Anhydrous hydrogen fluoride production.
- Elementary fluorine production.
- Uranium hexafluoride production.

To produce UF_6, high-temperature fluorination of uranium raw materials from fluorine gas is used. The elementary fluorine gas is produced by electrolysis of HF in the potassium dihydrotrifluoride melt. Under normal conditions the process system pressure does not exceed the atmospheric pressure.

Before release into the atmosphere, the process gases undergo special cleaning, leading to the formation of solutions that contain uranium and fluorine compounds. The solutions are transmitted to the liquid-wastes processing section in order to recover uranium.

To produce UF_6, both natural uranium concentrate, oxides or tetrafluoride, and slightly irradiated (the specific activity less than 22,000 Bq/g or 0.6×10^3 Ci/kg) uranium oxide may be used as raw materials.

All of the natural uranium raw materials are supplied by uranium ore mining and processing enterprises, both Russian and foreign. The oxide of slightly irradiated uranium is produced and delivered by SCC.

Cessation of HEU production resulted in the emergence of excessive enrichment capacities. This, together with the low cost of separation work due to the transition of the enrichment plants to centrifugal technology, allowed the plants to use depleted uranium (former years' "tailings" of the enrichment plants with ^{235}U content equal to 0.26–0.30%) as feed material.

In these circumstances, the utilization of the primary UF_6 production plants has fallen to between 10% and 15% of their capacity, whereas the enrichment plants' utilization is kept at the nominal level.

The potential risk of UF_6 production and uranium enrichment works is connected with the following factors:

- Nuclear risk from handling of enriched uranium compounds and probability of a self-sustained chain reaction (SCR) of uranium nuclei fission. It should be emphasized that there is a big difference in risk between production of LEU and production of, e.g., 95% HEU.
- Radiation hazard of uranium decay products while handling uranium compounds.
- Chemical hazard of UF_6 and fluorine compounds during the production cycle.
- General industrial risk, which may be aggravated under certain conditions by the above-noted factors.

According to estimates that take into account the nature and peculiarities of UF_6 production and uranium enrichment process technology, probable (hypothetical) accidents (conditioned by emerging SCR) resulting in the release of UF_6 or other uranium compounds will not have catastrophic consequences and may be characterized as local and limited to production areas.

2.2 Characteristics of Accumulated Radioactive Materials and Radioactive Wastes in the Industrial Complex as a Whole and at Separate Enterprises

UF_6 production plants and uranium enrichment plants are the most environmentally clean links of the nuclear fuel cycle. The main type of accumulated radioactive materials (RMs) are tailings of uranium enrichment plants in the form of UF_6, which are kept in hermetic steel containers at special storehouses under continuous radiological control. As noted above, the previously accumulated RMs are being used now as a feed material for the production of enriched uranium.

Both liquid radioactive wastes (LRWs) and solid radioactive wastes (SRWs) are generated during UF_6 production and the operation of uranium enrichment plants. The LRWs are managed so that they are not accumulated in or released into open water basins. The sewage waters,

Table 2.1. Amounts and radioactivity of accumulated SRWs, as of 1 January 1997.

Enterprise	Production type	Amount of RW, t	Total activity, Bq
SCC	UF_6 production	25,490	9.06×10^{12}
	Uranium enrichment	10,610	1.40×10^{11}
AECC[a]	UF_6 production	1,500	6.43×10^{13}
UECC	Uranium enrichment	29,070	1.08×10^{12}
ECP	Uranium enrichment	6,680	8.51×10^{10}

[a]Only medium-specific activity RW.

Table 2.2. Amounts and radioactivity of accumulated radioactive sludges formed as a result of LRW decontamination, as of 1 January 1997.

Enterprise	Production type	Amount of sludges, m^3	Total activity, Bq
UECC	Uranium enrichment	30,194	3.33×10^{11}
ECP	Uranium enrichment	7,140	1.48×10^{10}

containing radionuclides with concentrations lower than PC_B, are not included in the LRW category.[1]

Data on amounts of the accumulated RWs and their activity from start up of production work until 1 January 1997 are presented in *Tables 2.1* and *2.2*. The SRWs are cakes, cinders, scrap, rubber articles, polyethylene, cotton articles, building garbage, and sludges formed as a result of remelting decontaminated metallic equipment or decontamination of LRW. To decrease accumulation of SRWs and metals, facilities were created at the enterprises in order to recycle decontaminated and salvaged equipment, which necessitates the partial remelting of decontaminated components.

Sludges containing residual activity and formed as a result of remelting are sent to SRW storage and disposal facilities. The greatest part of the equipment, decontaminated to levels meeting sanitary specifications, is sent to civilian enterprises as scrap to be reprocessed.

The scheme of LRW management described below gives information about the origin of radioactive sludges.

The volumes of generated LRW are insignificant; at the enterprises, there are no special wastewater disposal systems that allow for continuous

[1]Here and throughout the chapter, the comparison is given with standards set by NRB–76/87 and OSP–72/87 regulations.

collection and transportation of LRW for processing. All trap water containing uranium is collected in special reservoirs, periodically transferred to transportation vessels, and sent for processing (uranium recovery) in special subdivisions located at the industrial sites of the enterprises.

After the uranium recovery to a residual concentration that is not higher than 2 mg/l, the solutions are treated with lime milk and transferred in the form of lime slurry with uranium concentration less than 0.5 mg/l to slime-settling basins, located in specially allotted and designed areas at the enterprises.

The actual concentration of radionuclides (uranium) in the liquid phase (clarified water) of the slime tanks is much lower (by 20–25 times) than the PC_B established for drinking water.

As the slime-settling basins are filled with solid material, they are put in dead storage by covering their surface with soil. The actual radiation dose rate at 1 m from the covering of soil surface is approximately 0.10 mrem/hr as compared with the permissible value of 2.8 mrem/hr (0.028 mSv/hr).

Some data related to UF_6 production and uranium enrichment plants at the SCC are missing from the tables because, besides UF_6 production and the uranium enrichment plants, the SCC includes a number of other plants, and the scheme of LRW and SRW management and sewage water removal to open water reservoirs is designed so that it is difficult to separate the radiation effects of the individual plants on the environment (the summary data for the SCC are presented in Chapter 7).

The RW activity depends on the content of uranium isotopes and the products of their radioactive decay. The nuclide composition is determined by the type of raw material used: It may be either natural or slightly irradiated uranium.

According to radiation control data, for the complex as a whole, the following nuclides are registered: ^{234}U, ^{235}U, ^{236}U, and ^{238}U; insignificant amounts of ^{238}Pu, ^{239}Pu, etc.; and ^{90}Sr, ^{106}Ru, ^{137}Cs, and ^{144}Ce.

The radionuclides released into the atmosphere in 1996 are presented in *Table 2.3*. The values for all the enterprises are insignificant, either absolutely or compared with calculated standard values of maximum permissible releases (MPR).

A similar situation exists with radionuclide releases into open water reservoirs (*Table 2.4*).

Table 2.3. Radionuclide[a] releases into the atmosphere in 1996.

Enterprise	Release Absolute, Bq/yr	% of MPR
UF_6 production plants		
1. SCC	7.25×10^8	0.6
2. AECC	1.76×10^8	0.05
Total	9.01×10^8	–
Uranium enrichment plants		
1. SCC	1.48×10^7	0.006
2. AECC	1.81×10^8	0.05
3. UECC	4.00×10^9	1.0
4. ECP	7.40×10^7	0.1
Total	4.23×10^9	–

[a]The sum of α-active isotopes.

Table 2.4. Radionuclide releases into open water reservoirs in 1996.

Enterprise	Water reservoirs	Volume of sewage water, $10^3 m^3$	Radionuclide releases Bq/yr	Radionuclide releases % of MPR
AECC	Angara River	27,953	3.70×10^7	0.001
UECC	Neivo-Rudyansk water storage	13,718	2.58×10^{10}	16.7
SCC[a]	Tom River	3,532	–[a]	–[a]
ECP[b]	–[b]	–[b]	–[b]	–[b]

[a]The technological scheme of the sewage water collector makes it difficult to distinguish releases of individual enterprises of the nuclear complex.
[b]The design does not stipulate any release into open water reservoirs.

In 1996, the enterprises (excluding ECP) released a total of 45.2 million m^3 of sewage water, mainly equipment cooling water, and industrial sewage. The concentrations of radionuclides in the sewage waters are, as a rule, by two orders of magnitude lower than the drinking water standards, certified by NRB–76/87 regulations.

The siting and erection of the SRW disposal points are carried out in accordance with established orders and are coordinated by regulatory bodies. Like the construction of enterprises, the removal of the SRW disposal points from service and their placement in prolonged storage are carried out under projects worked out by special institutions to meet sanitary standards as presented in "Sanitary Rules for Radioactive Waste Management"

(SPORO-85). The SRW isolation and disposal are completed in accordance with licenses, which are given by regulatory bodies.

At the enterprises of the uranium enrichment industrial complex, three types of storage facilities are used for SRW storage and disposal.

1. Points designated for accumulation and disposal of low-level (first group of activity) wastes take the form of open rectangular pits or trenches. Their bottom and slopes are covered with a waterproof clay layer 0.5–1.0 m thick, preventing radionuclide penetration into the environment. The wastes are dumped loose in pits and trenches, or in kraft-paper and plastic bags, to a level about 0.5 m below the upper edge. The cavity is then filled with soil, and the soil layer is compacted to the ground level. To protect the areas of the SRW disposal points against flooding, drainage canals are built, and access roads are covered with asphalt. To watch for possible contamination of groundwater by radionuclides, observation wells for sampling the waters are drilled around the disposal points areas and along perimeters of the pits and trenches.
2. Storage and disposal of low-level (second group of activity) sludges, formed as a result of LRW decontamination, are carried out in special structures (slime collectors) – namely, banked-up pits filled with loamy soils. Bituminous concrete is poured on sand and gravel beds or asphalt and covered by a bitumen layer. A protective soil layer is then placed on the bottom and slopes of the pits. Lime suspension (pulp) is pumped to the pits by slime pipelines. Control for probable migration of radionuclides is conducted by groundwater sampling from observation boreholes arranged along the slime collectors' perimeter.
3. Isolation and disposal of medium-level SRW (second group of activity) are carried out in storage structures in the form of reinforced concrete waterproof reservoirs (bunkers), lined with steel sheets that provide corrosion resistance. The bunkers' design excludes ingress of atmospheric precipitation into the reservoirs and prevents radionuclide migration into the environment. Around the trench burials, there is a network of observation boreholes to monitor the underground waters' composition.

Specially equipped motor cars are used for SRW transportation.

The radiation situation in the areas of RW storage sites can be evaluated in terms of uranium nuclide content in environmental objects. The monitoring data for 1996 are as follows:

Object	Uranium nuclides content
Groundwater	8.14×10^{-2}–1.11^{-1} Bq/l
Soil	4.81 Bq/kg
Grass	1.48 Bq/kg
Snow	3.7×10^{-2} Bq/l
Air	3.7×10^{-6} Bq/l

All the sites of RW storage are located within the industrial areas of the enterprises. They differ greatly in their volumes, filling level, time of commissioning, and service period. The mean filling level is 98% for trench-type burials, 50% for slime collectors, and 58% for bunker-type burials.

Radiation monitoring data, including control for ^{90}Sr and ^{137}Cs radionuclides content in soil as well as gamma-radiation exposure rates, show that no territories were contaminated as a result of operation of the uranium enrichment industrial complex enterprises. Background levels of gamma-radiation exposure rates were recorded for many years at the territories of sanitary-protection zones (SPZs) and observation zones of the enterprises.

In 1996, the gamma-radiation background exposure rates were as follows:

SCC	6–20 μR/hr
AECC	10–22 μR/hr
UECC	10–20 μR/hr
ECP	10–20 μR/hr

Measurements of ^{90}Sr and ^{137}Cs content in soil made in 1996 at SCC may serve as evidence of the insignificant effect of uranium enrichment complex enterprises on the environment. According to the radiation monitoring data at the location of the UF_6 production and uranium enrichment plants, the radionuclide content in the soil does not exceed 1.5 kBq/m^2 for ^{90}Sr and 11 kBq/m^2 for ^{137}Cs in the SPZ of the enterprise, and does not exceed 0.04–1.1 kBq/m^2 and 0.4–4.8 kBq/m^2 in the observation zone for ^{90}Sr and ^{137}Cs, respectively.

Bibliography

Energoatomizdat, 1986, Sanitary Rules for Radioactive Waste Management (SPORO-85), Energoatomizdat, Moscow, Russia.

Energoatomizdat, 1988a, Radiation Safety Standards (NRB- 76187), Energoatomizdat, Moscow, Russia.

Energoatomizdat, 1988b, Basic Sanitary Rules (OSP-72/87), Energoatomizdat, Moscow, Russia.

Kruglov, A.K., 1994, *How the Nuclear Industry Was Created in the USSR*, p. 380, TsNIIAtominform, Moscow, Russia.

Minatom (Ministry of the Russian Federation for Atomic Energy), 1995, Analytical information of enterprises on environmental effects of UF_6 production and uranium enrichment plants, Minatom, Moscow, Russia.

Chapter 3

Nuclear Fuel Production

3.1 Introduction

Before 1991, industrial production of nuclear fuel was carried out only at the enterprises in the Russian Federation (RF). At present these enterprises comprise the Joint Stock Society (AO) TVEL:

- AO Machinery Plant, Electrostal, Moscow region.
- AO Chepetsky Mechanical Plant, Glazov, Udmurt Republic.
- AO Novosibirsk Plant of Chemical Concentrates, Novosibirsk region.
- State Enterprise Moscow Plant of Polymetals, Moscow.

The Production Amalgamation (PO) Ulbinsky Metallurgical Plant, in Ust-Kamenogorsk, Republic of Kazakhstan, in cooperation with the enterprises of the AO TVEL, has also participated in the production of nuclear fuel.

Uranium ores and concentrates, uranium oxides, and hexafluoride of natural uranium or uranium enriched by ^{235}U are the source materials for the production of nuclear fuel.

Metallic uranium, its alloys, uranium dioxide fuel of various degrees of enrichment by ^{235}U, and fuel compositions are produced during the chemical and metallurgical processes. The end products are:

- Fuel elements and fuel assemblies for nuclear reactors of various applications.
- Aluminum-clad natural uranium fuel for plutonium production.

- Slugs of enriched uranium.
- Fuel elements and fuel assemblies for transport nuclear installations (high enrichment by ^{235}U – up to 95%).
- Fuel elements and fuel assemblies of various degrees of enrichment by ^{235}U for research reactors.

Because radioactive sludges and solids are created as a by-product of the nuclear materials production process, the enterprises must properly dispose of wastes in storage facilities. Disposal of radionuclides into watersheds and into the atmosphere at all the enterprises is at levels lower than standards allow, and there have been no radiation accidents leading to the contamination of the environment.

3.2 Characteristics of Radioactive Wastes

The data on radioactive wastes (RWs) accumulated at the enterprises for the production of nuclear fuel as of 1 January 1996 are presented in *Table 3.1*. At the enterprises in the RF, there are solid low- and medium-active wastes that contain uranium radionuclides and ^{226}Ra. Their total quantity is 5,650,000 tons, with an activity of 1.4×10^{14} Bq.

According to the inventory of RWs as of 1 January 1990, there are liquid low-active wastes at the PO Ulbinsky Metallurgical Plant with a volume of 939,000 m^3, and they contain uranium radionuclides with an activity of 2.3×10^{11} Bq. Additionally, there are 1,351,000 tons of solid radioactive wastes (SRWs) at the enterprise, which are low- and medium-level RWs with an overall activity of 3.8×10^{13} Bq. This activity is determined by uranium radionuclides and by spent sealed sources of ionizing radiation containing a wide variety of nuclides (^{241}Am, ^{90}Sr, ^{239}Pu, ^{60}Co, ^{170}Tm, ^{124}Sb, etc.).

Table 3.2 shows the following characteristics of the storage facility of RWs situated at nuclear fuel production enterprises: period of operation, area, volume of storage facility, total activity, and level of waste activity in storage facility. At this time, some storage facilities have already been filled. They are currently under observation and maintenance.

Table 3.3 shows the area of territories contaminated by radionuclides (with special attention given to the corresponding territories on the industrial sites of the enterprises), size of the sanitary-protection zones (SPZs), and area of the zones of supervision.

Table 3.1. Characteristics of RWs accumulated at nuclear fuel production enterprises as of 1 January 1996.

Enterprise	Activity, Bq	Main contaminating radionuclides
Russian Federation		
1. AO Machinery Plant	4.2×10^{13}	Uranium radionuclides ^{226}Ra
2. AO Chepetsky Mechanical Plant	7.0×10^{13}	Uranium radionuclides ^{226}Ra
3. AO Novosibirsk Plant of Chemical Concentrates	3.0×10^{13}	Uranium radionuclides ^{226}Ra
4. State Enterprise Moscow Plant of Polymetals	n.a.	n.a.
Total	1.42×10^{14}	
Republic of Kazakhstan		
PO Ulbinsky Metallurgical Plant (as of 1 January 1990)	3.8×10^{13}	Uranium radionuclides Emitters ^{241}Am ^{90}Sr ^{239}Pu

n.a. = not available

The overall area of the enterprises contaminated by radionuclides in the RF as of 1 January 1996 is 1.7 km^2, of which 0.2 km^2 has a dose rate of more than 200 μR/hr. The larger part of the contaminated areas is located at the industrial sites of the enterprises (1.5 km^2). The SPZs have 0.11 km^2 of such areas, and in the zones of supervision, 0.12 km^2 have a dose rate of more than 200 μR/hr. Data as of 1 January 1990 show that the PO Ulbinsky Metallurgical Plant has an area of 0.6 km^2 in the SPZ that is contaminated, including 0.07 km^2 with a dose rate of more than 200 μR/hr.

At all the enterprises, the contamination is primarily due to uranium radionuclides.

3.3 Evaluation of the Completeness of Information

Data given in *Tables 3.1–3.3* satisfactorily show quantitative characteristics of the presence of RWs and their activity. Recently (as of 1 January 1990 and 1 January 1993), an inventory was carried out of places and objects of extraction, transportation, processing, use, collection, storage facility, and burial of RWs and sources of ionizing radiation.

Table 3.2. Storage of RWs at nuclear fuel production enterprises, as of 1 January 1996.

Enterprise	Description of storage facilities	Period of operation	Area, $10^3 m^2$	Volume, $10^3 m^3$ Planned	Volume, $10^3 m^3$ Actual	Total activity, Bq	Level of RW activity
Russian Federation							
1. AO Machinery Plant	Storage No. 240.	1949–1951	9.4	–	16	1.0×10^{13}	Low- and medium-level
	Storage No. 298 has trenches without filtration-proof barriers, filled with nonradioactive wastes.	1950–1981	93.75	–	135	3.2×10^{13}	Low- and medium-level
	Storage No. 294a is a surface-type facility with filtration-proof layer.	1981–present	30.1	120	104	9.2×10^{10}	Low-level
2. AO Chepetsky Mechanical Plant	Old storage is now under conservation.	1952–1980	425	–	2,800	1.5×10^{13}	Low-level
	Presently operating facility is a plain, poured type, without filtration-proof protection.	1965–present	450	3,600	2,900	5.5×10^{13}	Low-level
3. AO Novosibirsk Plant of Chemical Concentrates	The first section of the storage facility is now under conservation.	1954–1964	42	–	55	7.4×10^{12}	Low-level
	The second section of the facility is a ravine, poured type with a single dam without filtration-proof protection.	1964–present	160	950	800	2.3×10^{13}	Low-level
4. State Enterprise Moscow Plant of Polymetals	n.a.	1932–1972	–	–	126	–	Low- and medium-level

Table 3.2. Continued.

Enterprise	Description of storage facilities	Period of operation	Area, $10^3 m^2$	Volume, $10^3 m^3$		Total activity, Bq	Level of RW activity
				Planned	Actual		
Republic of Kazakhstan							
PO Ulbinsky Metallurgical Plant (as of 1 January 1990)	Old storage for solid wastes is without filtration-proof barrier.	1953–1970	64	500	500	1.0×10^{13}	Low- and medium-level
	Plot 1c of the new storage is of poured type with bottom condensed with clay and solid RWs on top.	1962–1986	73.5	400	400	4.5×10^{12}	Low- and medium-level
	Plots 1 and 3 are of poured type with bottom condensed with clay.	1970–1986	227	1,700	1,700	9.4×10^{10}	Low-level
	Plot 2 is of poured type with bottom condensed with clay and polyethylene layer on top.	1987–present	251	2,500	620	1.3×10^{11}	Low-level
	Burial facility for solid wastes has stainless steel coating.	1969–1970	0.48	2	2	1.0×10^{12}	Medium-level
	Burial facility for spent sealed sources of ionizing radiation has concrete tank with sump of stainless steel.	1976–present	0.02	0.24	0.1	2.2×10^{13}	Spent sources

Table 3.3. Territories contaminated with radionuclides as a result of activities at nuclear fuel production enterprises, km^2.

Enterprise	Total contaminated territory	Contaminated territory of the industrial site			Contaminated territory in the SPZ			Contaminated territory in the zone of supervision		
			With dose rate			With dose rate			With dose rate	
		Total	$\leq$ 200 μR/hr	$\geq$ 200 μR/hr	Total	$\leq$ 200 μR/hr	$\geq$ 200 μR/hr	Total	$\leq$ 200 μR/hr	$\geq$ 200 μR/hr
Russian Federation										
1. AO Machinery Plant	0.262	0.141	–	0.141	0.001	0.001	–	0.12	–	0.12
2. AO Chepetsky Mechanical Plant	1.352	1.342	1.283	0.059	0.010	0.007	0.003	–	–	–
3. AO Novosibirsk Plant of Chemical Concentrates	0.048	0.001	–	0.001	0.047	–	0.047	–	–	–
4. State Enterprise Moscow Plant of Polymetals	0.059	0.004	0.004	0.0003	0.055	0.055	0.0001	–	–	–
Total	1.721	1.488	1.287	0.201	0.113	0.063	0.050	0.12	–	0.12
Republic of Kazakhstan										
PO Ulbinsky Metallurgical Plant (as of 1 January 1990)	0.556	n.a.	n.a.	n.a.	0.556	0.483	0.073	n.a.	n.a.	n.a.

n.a. = not available.

The enterprises report annual data on the disposals into water and atmospheric releases of radioactive substances, as well as data on the discovery and rehabilitation of the contaminated territories to the governing body. Databases on disposals into water, atmospheric releases, and contaminated territories of the enterprises of the AO TVEL date back to 1 January 1993. Data on the PO Ulbinsky Metallurgical Plant situated in the Republic of Kazakhstan are limited by the materials inventoried as of 1 January 1990.

A comprehensive study of the contamination of territories, the state of storage facilities, the migration of radionuclides from them, and the influences on the population living in the territories adjacent to the enterprises is advisable to determine the need of rehabilitation of the contaminated territories and to complete engineering and research work. "The Rehabilitation of Territories Contaminated as a Result of the Activity of the Enterprises for Nuclear Materials Production for Nuclear Weapons" program outlines the work on the rehabilitation of contaminated territories at the AO Machinery Plant, AO Chepetsky Mechanical Plant, AO Novosibirsk Plant of Chemical Concentrates, and the State Enterprise Moscow Plant of Polymetals.

The *AO Machinery Plant* is situated in the city of Electrostal, in the Moscow region. It was the first enterprise in the Soviet Union to produce industrial metallic uranium and nuclear fuel. The work with uranium began in 1946.

Uranium-containing ores and uranium concentrates were processed at the plant to produce pure uranium salts and metallic uranium. They were subsequently cast, underwent plastic deformation, were mechanically treated (to obtain molds of the necessary configuration), and aluminum clad, in order to produce standard uranium slugs. Uranium ore processing ended in 1957, and manufacture of standard uranium slugs ended in 1967.

In the 1950s, the production of uranium slugs enriched by ^{235}U began. In subsequent years, the production of fuel elements, fuel assemblies for various types of research reactors, nuclear reactors for nuclear power engineering, and transport facilities was begun. Fuel elements were made of enriched uranium in the form of uranium dioxide or other compositions. All of the above-mentioned items are still produced.

RWs that have resulted from uranium ore processing and nuclear fuel production are concentrated in two surface storage facilities (No. 240 and

No. 298) that are not used now. These storage facilities are situated on the territory of the enterprise, and are in the form of trenches without filtration-proof barriers. After natural liquid evaporation and filtration, the trenches are filled with non-RWs, outdated machinery, and metallic scrap. The storage facilities, which have a filled volume of 151,000 m^3, are in need of rehabilitation. No. 240 was used from 1949 until 1951, and No. 298 was used from 1950 until 1981. The overall activity of the two storage facilities is 4.2×10^{13} Bq and is primarily radionuclides of ^{235}U and ^{238}U as well as ^{226}Ra. There are no liquid radioactive wastes (LRWs) at the storage facilities.

Since 1981, RWs in the form of sludge processed by liming have been sent to surface sludge storage facility No. 294a. Its bottom and walls are protected by filtration-proof clay material. The planned volume of the storage facility is 120,000 m^3, and to date, the facility has been filled with 104,000 m^3 of waste (including water). In the liquid phase, activity is within the permissible concentration. SRWs have a volume of 5,000 m^3 with an overall activity of 9.2×10^{10} Bq. These are low-active wastes containing uranium radionuclides whose specific activity is about 1.8×10^4 Bq/kg. Radionuclides placed into the storage facility during 1995 were 7.4×10^9 Bq.

Disposal of sewage waters at the enterprise takes place through several outlets. The concentration of alpha-active radionuclides in them is lower than 0.4 Bq/l (with the exception of outlet No. 5); the permissible concentration is 81.4 Bq/l. Before 1976, waters contaminated with radionuclides were disposed through outlet No. 5, which resulted in the accumulation of uranium and radium in silts. Currently, radionuclides are gradually being leached from the silts. In 1995, the content of alpha-active radionuclides discharged from outlet No. 5 was 6.3 Bq/l (2×10^9 Bq for the year), which is less than 10% of the standard disposal. In 1995, atmospheric release of RWs (uranium radionuclides) at the enterprise was 6.3×10^8 Bq, which is 68% of the permitted release.

The area of the industrial site contaminated by uranium radionuclides and radionuclides of ^{226}Ra is equal to 0.14 km^2, due to plots having local contamination and containing surface storage facilities of RWs with dose rates of more than 1 mR/hr. In the SPZ and in the zone of supervision, the contaminated area is 0.12 km^2, due to parts of the B2 brook and water basin in the Fryazevo state farm, which contains contaminated silts with dose rates from 0.2–1 mR/hr.

The *AO Chepetsky Mechanical Plant* is situated in the city of Glazov, Udmurt Republic. Work with uranium began in 1949. The source material was uranium concentrates, including those depleted in the radionuclide ^{235}U, and uranium-containing ores that were processed with the result of obtaining uranium tetrafluoride and metallic uranium. Standard uranium slugs for industrial reactors were made of metallic uranium. The production of standard slugs started to diminish beginning in 1986. The production was transferred to the Novosibirsk Plant of chemical concentrates where metallic uranium is sent from the Chepetsky Mechanical Plant.

LRWs from uranium production in the form of sludge (after they are processed with lime) are sent through the pipeline to a plain poured-type storage facility with a protective dam made of a sand-and-gravel mixture and no filtration-proof barrier. The storage facility, which was used between 1952 and 1980, is now under conservation. Its volume is 2.8 million m^3, and its overall activity is 1.5×10^{13} Bq (low-level RWs). Radioactivity is due to radionuclides of uranium, thorium, and ^{226}Ra. The present storage facility for RWs is similar to the above-mentioned one. It has no filtration-proof barrier and has been used since 1965. The liquid settled in it has a concentration of uranium lower than 1.8 mg/l, and, therefore, can be treated as non-RW. The liquid is sent to the underground storage facility and discharged at a depth of 1,500 m. The capacity of the storage facility is 3.6 million m^3 and is 80% filled. The overall activity of the wastes in the storage facility is 5.5×10^{13} Bq.

SRWs are transported to the site for waste processing where flammable wastes are burned, outdated machinery is cut, and non-flammable solid wastes are stored. The overall amount of RWs accumulated at the enterprise is 4 million tons.

Currently, the concentration of radionuclides in the sewage waters disposed into the Cheptsa River does not exceed the maximum permissible value. These sewage waters include cooling waters of the thermoelectric station and storm runoff, clarified disposals from ash storage facility of the thermoelectric station, and common disposals after their biological decontamination. In 1995, the overall activity of radionuclides (uranium, thorium, ^{226}Ra, and ^{210}Po) in sewage waters (54 million m^3) was equal to 1.2×10^{11} Bq. The uranium radionuclides amount to 17% of permissible concentration.

Atmospheric releases of uranium radionuclides in 1995 were 9.6×10^8 Bq – that is, 74% of maximum permissible value.

As a result of the activity of the enterprise, the industrial site and the SPZ contain lands contaminated by radionuclides of uranium, thorium, and ^{226}Ra. These lands include storage facilities (both filled and operating), industrial sites where ores had been heaped in the first years of the enterprise's operation, territories of the old enriching plant and of the now-operating uranium hydrometallurgical technological process, and the first stage of the ash storage facility for the thermoelectric station, which had been previously used for storage of contaminated machinery. The contaminated area on the industrial site is 1.34 km^2, with dose rates from 20–200 μR/hr. Some plots have dose rates of more than 1 mR/hr. In the SPZ, the area of the contaminated plots is 0.01 km^2, with dose rates of up to 1 mR/hr. Specific alpha activity of the contaminated territories lies in the interval $(7.4\text{–}74)\times 10^3$ Bq/kg.

The whole complex of measures aimed at the rehabilitation of the contaminated territories has been worked out at the enterprise. A new storage facility equipped with a filtration-proof barrier is planned, as well as a special facility for the processing of SRWs.

The *AO Novosibirsk Plant of Chemical Concentrates* is situated at the outskirts of the city of Novosibirsk. The work with uranium began here in 1949. The source materials were uranium ore and concentrates (before 1978) and uranium hexafluoride of various degree of enrichment by ^{235}U. Chemical processing with the aim of obtaining uranium oxides and metallic uranium is being carried out. The Chepetsky Mechanical Plant also supplies the enterprise with metallic uranium, and the PO Ulbinsky Metallurgical Plant supplies the plant with pellets of uranium dioxide. The end products are standard slugs of natural uranium and fuel assemblies of enriched uranium for various purposes.

RWs containing uranium radionuclides are transported by the sludge pipeline to the storage facility that is situated 7.5 km from the city. The storage facility is a ravine, poured type with a single dam.

The first stage of the storage facility operated from 1954 to 1964 and had no filtration-proof barriers. At present, the sludge at the storage facility is partly dried and is overgrown with vegetation. The area of the storage facility is 0.042 km^2, and the amount of solid wastes is 360,000 tons with an activity of 7.4×10^{12} Bq.

The second stage of the storage facility has been operating since 1964. Its planned size is 0.16 km^2, and the planned volume is 950,000 m^3. Currently, 800,000 m^3 are filled in, with an activity of 2.3×10^{13} Bq. As a result of the filtration of the water, 50,000 m^3 of water runs out from the lowest part of the storage facility annually. A part of the water is pumped into the storage facility again and the remaining part flows into nearby rivers and brooks. Taking into account the critical state of the dam of the second stage of the storage facility, the lowest part of the dam was reinforced with sand, road metal, and other materials to make it higher.

The third stage of the storage facility is under development. It will have a clay barrier. Clarified water will be returned to the enterprise.

The overall amount of SRWs, as of 1 January 1996, is 1,270,000 tons. The specific activity is 2.4×10^4 Bq/kg.

Sewage waters containing radionuclides are not transported into open water reservoirs.

In 1995, atmospheric releases of ^{238}U were 4.8×10^8 Bq and that of ^{235}U were 5.2×10^8 Bq, which amounts to 60% and 80% of permitted release, respectively.

The industrial site and the SPZ of the enterprise contain lands contaminated with uranium radionuclides, including the first and second stages of the storage facility, the territory where the sludge pipeline connects the industrial site and storage facility areas, the floodplain of the Eltsovka River, and other sites. The size of the contaminated territory of the industrial site is 0.001 km^2, and in the SPZ the contaminated area is 0.047 km^2, with dose rates of up to 1 mR/hr. Rehabilitation measures are underway on the contaminated territories of the enterprise.

The *State Enterprise Moscow Plant of Polymetals* is situated in Moscow on the bank of the Moskva River and is surrounded by highly populated urban areas. From 1932 to 1972 the plant processed radioactive source material containing natural radionuclides of thorium and uranium and produced thorium and high-purity rare earth products. At present, the plant does not work with radioactive substances. Therefore, there are neither discharge of radionuclides into water nor atmospheric releases.

As of 1 January 1996, 0.059 km^2 of land was contaminated, including 0.004 km^2 on the industrial site (separate spots) and 0.055 km^2 in the SPZ (the bank of the Moskva River). Dose rates reach 1 mR/hr, and radioactivity is determined mainly by radionuclides of thorium. The con taminated soils along the banks of the Moskva River must be treated as SRWs (low and medium levels) and plots of radioactive contamination themselves must be treated as uncontrolled storage. The overall amount of the contaminated soils exceeds 126,000 m^3. Given the possibility of landslides of the bank, movement of radioactive substances into water cannot be excluded. Measures against landslides have been taken on the steep bank of the river. Slow migration of radionuclides in the direction of the river is, however, still observed.

Data on the radiation situation at the enterprise show that first-priority measures should be taken to guarantee the ecological safety of the nearby population. These include maintenance of the abnormal plots of the steep bank to prevent further contamination of the Moskva River and removal of the surface sewer from the steep bank to prevent dispersion of radionuclides and washout of the contaminated soils. This work is presently being carried out.

The *PO Ulbinsky Metallurgical Plant* is in the city of Ust-Kamenogorsk, Republic of Kazakhstan. The first radioactive material – thorium oxalate – was produced at this enterprise in 1949. Since 1953, concentrates of uranium have been processed to get oxides. In 1973, the production of fuel pellets from dioxide uranium powder enriched by ^{235}U was organized. The end product, which is designed for the manufacture of fuel elements of energy reactors, is currently transported to the Novosibirsk Plant of Chemical Concentrates.

RWs containing radionuclides of uranium are transported to storage. The data, as of 1 January 1990, are given below. The old storage facility areas were in use between 1953 and 1970. They are situated 1 km from the industrial site and measure 0.064 km^2. The bottoms of the facilities are compacted with loam and plastic clay. Local soil was used to cover the storage facility and the surface is now overgrown with grass. The storage facilities contain solid wastes from various processes, including 350,000 tons of uranium production wastes with medium-level RWs (10^{13} Bq). Gamma radiation at the storage site does not exceed the background level – that is, 14–17 μR/hr. The new storage facility is situated 2.3 km from the enterprise.

Plot 1c, which operated between 1962 and 1986, has a bottom compacted with clay and measures 0.073 km^2. It was filled with non-RWs from beryllium production. After cessation, 1 million tons of gypsum fluoride (containing radionuclides of the uranium family with a medium-level activity of 4.5×10^{12} Bq) were put on the surface of the storage facility in order to seal it. Dose rates on the site of the storage facility are 60–1,000 μR/hr.

The plot of shops Nos. 1 and 3 operated between 1970 and 1986. The bottom is covered with clay and the size is 0.227 km^2, with a planned volume of 1.7 million m^3. Clarified water is not returned to the enterprise. The sewage of various production works was stored in bulk, including 320,000 m^3 of RWs from uranium production with an activity of 9.4×10^{10} Bq. After reaching design capacity, clarified sewage was transported to plot 1 or 2. Dose rates are up to 200 μR/hr.

Plot 2 has been operating since 1987. It is a poured-type plot with compacted soil, clay closure, and polyethylene layer. Its size is 0.251 km^2, and its planned volume is 2.5 million m^3. Some 620,000 m^3 of the storage facility are filled in with low-active sludge from uranium production, and its activity is 1.3×10^{11} Bq. The dose rate is up to 200 μR/hr. The wastes from beryllium and tantalum production are also transported to plot 2.

The construction of a new plot with a volume of 4.3 million m^3 and size of 0.32 km^2 is planned.

In 1969–1970, a burial storage facility with a coating of stainless steel was operating at the enterprise. It contains 445 tons of solid wastes of uranium production with medium-level activity of 1,012 Bq. The dose rate at the storage facility is up to 60 μR/hr.

In 1976, a burial facility with a sump of stainless steel for spent sealed sources of ionizing radiation was put into operation. Its volume is 24 m^3. As of 1 January 1990, the following emitters were stored there: long-lived isotopes of ^{241}Am, 4.4×10^{11} Bq; ^{90}Sr, 1.85×10^{10} Bq; ^{239}Pu, ^{60}Co, ^{147}Pm, ^{204}Tl, less than 10^8 Bq; and relatively short-lived isotopes of ^{170}Tm, 2.2×10^{13} Bq; ^{124}Sb, 1.1×10^{11} Bq, etc.

As of 1 January 1990, the following amounts of wastes were at the enterprise: SRWs, 135,100 tons with an activity of 3.8×10^{13} Bq; LRWs, 939,000 m^3 with an activity of 2.3×10^{11} Bq.

In 1990, the content of uranium radionuclides in the sewage waters of the enterprise was equal to 4.4×10^9 Bq, which is less than permitted levels.

Atmospheric releases in 1990 were equal to 5.2×10^9 Bq of uranium radionuclides – that is, 80% of permitted release.

The overall size of the territory contaminated with radionuclides as of 1 January 1990 was determined by the storage facilities situated in the SPZ of the enterprise and was equal to 0.6 km^2, with a dose rate exceeding 200 μR/hr.

Bibliography

Energoatomizdat, 1995, *The Production of the First Soviet Atomic Bomb* (Collection of articles), pp. 316–332, Energoatomizdat, Moscow, Russia.

Golovin, S.V., Benkevich, V.P., Marchenko, S.M., and Farushev, I.A., 1996, *The Creation and Development of the AO Chepetsky Mechanical Plant*, Booklet No. 544, Izhevsk Magistrate, Izhevsk, Russia.

ISTC (International Science and Technology Center), 1996a, Phase 1: The creation of a simple operational database with connection to GIS, including presently available data on the radiation legacy of the ex-USSR, In: *The Final Technical Report on the ISTC Project No. 245* (Radleg), ISTC, Moscow, Russia.

ISTC (International Science and Technology Center), 1996b, Radleg simple operational database (tables 1, 2, 3, 10, 11, and 12), ISTC, Moscow, Russia.

Krugov, A.K., 1994, *How the Nuclear Industry Was Created in the USSR*, pp. 285–314, TsNIIAtominform, Moscow, Russia.

Minatom (Ministry of the Russian Federation for Atomic Energy), 1996, *Fifty Years of Russian Nuclear Science and Technology* (Collection of articles), pp. 164–173, Minatom, Moscow, Russia.

Minatom (Ministry of the Russian Federation for Atomic Energy) *et al.*, 1991, *Summary Report on Inventory of Radioactive Wastes, Their Storage and Burial Locations* (as of 1 January 1990), VNIICHT Funds, Moscow, Russia.

Minatom (Ministry of the Russian Federation for Atomic Energy) *et al.*, 1994, *Cadastre of Radioactive Wastes Storage and Burial Locations*, VNIICHT Funds, Moscow, Russia.

NPO Taifun, 1993, *Radiation Situation on the Territory of Russia and Nearby States in 1992* (Collection of articles), pp. 89–96, NPO Taifun, Obinsk, Russia.

NPO Taifun, 1995, *Radiation Situation on the Territory of Russia and Nearby States in 1994* (Collection of articles), pp. 106–113, 323-336, NPO Taifun, Obinsk, Russia.

Russian Federation Federal Purpose Program, 1995, *Handling of Radioactive Wastes and Spent Nuclear Materials: Their Utilization and Burial (1996–2005)* (Program approved by a resolution of the government of the Russian Federation on 23 October 1995, No. 1030), Russian Federation, Moscow, Russia.

Tomsk University, 1996, *Radioactivity and Radioactive Elements in the Human Environment* (Collection of articles), pp. 207–212, Tomsk University, Tomsk, Russia.

Chapter 4

Nuclear Power Plants

4.1 Introduction

By the time of the disintegration of the Soviet Union in 1991, the country's nuclear power had become an important part of the national electric power system, constituting 12.6% of total electrical output. In the largest electric power grids of the Northwest, Central, and Middle Volga, nuclear electricity generation was particularly important and produced 32%, 21.3%, and 13.8%, respectively, of total electricity output.

In nuclear power plant (NPP) operations, some amount of radioactive material (RM) is invariably released to the environment. In addition, NPP accidents are a potential source of radiation to NPP personnel, the general population, and the environment. Nuclear power reactor plants are designed so that the largest part of radioisotopes produced is isolated from the biosphere. Only a small amount of radionuclides is released to the environment with gaseous and aerosol discharges and liquid effluents; they dissipate into the atmosphere and surface water in the vicinity of the NPP. However, more than 40 years of experience with radiation monitoring of the Soviet/Russian environment shows that the discharge of such quantities of radionuclides to the biosphere has not resulted in changing the natural radiation background of the area surrounding the NPP site.

Before the disintegration of the Soviet Union in 1991, 17 NPPs were in operation with 51 power units and a total installed capacity of 38.2 GW. Included in this capacity are four units at the decommissioning stage – AMB–100 and AMB–200 units at Beloyarsk NPP, and VVER–210 and VVER–365 units at NovoVoronezh NPP – and two VVER–440 units at

Table 4.1. NPPs in the FSU (as of 1991).

Name (location)	Unit	Reactor type	Date of commissioning	Date of decommissioning	Thermal power, MW	Electric power, MW	Fuel type
Balakovo	1	VVER–1000	1985	2015	3,000	1,000	UO_2
(8 km from	2	VVER–1000	1987	2017	3,000	1,000	UO_2
Balakovo,	3	VVER–1000	1988	2018	3,000	1,000	UO_2
Saratov region)							
Beloyarsk	1	AMB–100	1964	1981	286	100	U+Mg
(40 km from	2	AMB–200	1967	1989	565	200	U+Mg
Yekaterinburg)	3	BN–600	1980	2010	1,470	600	U
Bilibino	1	EGP–6	1974	2004	60	12	U+Mg
(5 km from	2	EGP–6	1974	2004	60	12	U+Mg
Bilibino,	3	EGP–6	1975	2005	60	12	U+Mg
Magadan	4	EGP–6	1976	2006	60	12	U+Mg
region)							
Kalinin	1	VVER–1000	1984	2014	3,000	1,000	UO_2
(3 km from	2	VVER-1000	1986	2016	3,000	1,000	UO_2
Udomlya,							
Tver region)							
Kola	1	VVER–440	1973	2003	1,375	440	UO_2
(Murmansk	2	VVER-440	1974	2004	1,375	440	UO_2
region)	3	VVER-440	1981	2011	1,375	440	UO_2
	4	VVER-440	1984	2014	1,375	440	UO_2
Kursk	1	RBMK–1000	1976	2006	3,200	1,000	UO_2
(40 km from	2	RBMK–1000	1979	2009	3,200	1,000	UO_2
Kursk, Kursk	3	RBMK–1000	1983	2013	3,200	1,000	UO_2
region)	4	RBMK–1000	1985	2015	3,200	1,000	UO_2
Leningrad	1	RBMK–1000	1974	2005	3,200	1,000	UO_2
(90 km from	2	RBMK–1000	1975	2005	3,200	1,000	UO_2
St. Petersburg)	3	RBMK–1000	1979	2009	3,200	1,000	UO_2
	4	RBMK–1000	1981	2011	3,200	1,000	UO_2
NovoVoronezh	1	V–210	1964	1984	760	210	UO_2
(42 km from	2	V–365	1969	1990	1,322	365	UO_2
Voronezh)	3	V–440	1971	2001	1,375	416	UO_2
	4	V–440	1972	2002	1,375	416	UO_2
	5	V–1000	1980	2010	3,000	1,000	UO_2

Table 4.1. Continued.

Name (location)	Unit	Reactor type	Date of commissioning	Date of decommissioning	Thermal power, MW	Electric power, MW	Fuel type
Smolensk	1	RBMK–1000	1982	2012	3,200	1,000	UO_2
(40 km from	2	RBMK–1000	1985	2015	3,200	1,000	UO_2
Roslavl, Smolensk region)	3	RBMK–1000	1990	2020	3,200	1,000	UO_2
Armenia	1	VVER–440	1976	1989	1,376	407.5	UO_2
(15 km from Oktemberan)	2	VVER–440	1970	1989	1,375	407.5	UO_2
Zaporozhie	1	VVER–1000	1984	2014	3,000	1,000	UO_2
(40 km from	2	VVER–1000	1985	2015	3,000	1,000	UO_2
Kamenka,	3	VVER–1000	1986	2016	3,000	1,000	UO_2
Dnieperopet-	4	VVER–1000	1987	2017	3,000	1,000	UO_2
rovsk region)	5	VVER–1000	1989	2019	3,000	1,000	UO_2
Ignalina	1	RBMK–1500	1984	2014	4,800	1,500	UO_2
(41 km from Ignalina)	2	RBMK–1500	1987	2017	4,800	1,500	UO_2
Rovno	1	VVER–440	1980	2010	1,375	402	UO_2
(60 km from	2	VVER–440	1981	2011	1,375	416	UO_2
Sarny)	3	VVER–1000	1986	2016	3,000	1,000	UO_2
Khmelnitsk (30 km from Izyaslav)	1	VVER–1000	1987	2017	3,000	1,000	UO_2
Chernobyl	1	RBMK–1000	1977	2007	3,200	1,000	UO_2
(18 km from	2	RBMK–1000	1978	2008	3,200	1,000	UO_2
Chernobyl)	3	RBMK–1000	1981	2011	3,200	1,000	UO_2
	4	RBMK–1000	1983	1986	3,200	1,000	UO_2
South	1	VVER–1000	1982	2012	3,000	1,000	UO_2
Ukraine (35 km from Voznesensk)	2	VVER–1000	1985	2015	3,000	1,000	UO_2

the Armenian NPP that were shut down as a result of public protests exaggerated by the earthquake in 1989.

After the disintegration of the Soviet Union, nine NPPs with 28 power units with a total installed capacity of 20.24 GWe remained in Russia: five NPPs with 14 power units with a total installed capacity of 12.88 GWe are in Ukraine; Lithuania has the Ignalina NPP with two RBMK–1500 units; Armenia has two inoperative VVER–440 units; and Kazakhstan has one BN–350 power unit.

In view of the energy crisis in Armenia caused by political and economic problems, Unit–2 of the Armenian NPP was again put into operation in November 1995 with the assistance of Russian experts.

The information about NPP locations, reactor types, number of units, and their capacity in the former Soviet Union (FSU) is given in *Table 4.1*.

4.2 Characteristics of Radioactive Materials and Wastes Produced in Nuclear Power Plants

4.2.1 Radioactive material and waste sources at NPPs

In NPP operation, RM and wastes are stored in reactor storage facilities for spent nuclear fuel (cooling ponds), and in separate liquid and solid storage facilities at the NPP site.

Table 4.2 lists data on the amount of spent nuclear fuel (SNF) stored at NPP sites – including storage in cooling ponds and separate storage at NPPs with RBMK reactors – as well as the estimate of the total fission products (FP) activity in the spent fuel.

Because the spent fuel is periodically transported – for example, from a VVER–440 reactor to a reprocessing facility and from an NPP with a VVER–1000 reactor to a centralized repository – the quantity of the stored fuels at NPPs also changes. Therefore, in the input database, the amount of spent fuel at a given moment can be only estimated. Another radioactivity source at the NPP is solid and liquid wastes produced in the course of plant operation. The classification of liquid radioactive wastes (LRWs) and solid radioactive wastes (SRWs) adopted in Russia is based

Table 4.2. Data on the amount and activity of FPs in the SNF stored at NPP sites.

NPP	Data as of 1 January 1991: SNF mass, t	Data as of 1 January 1991: FP activity, EBq[a]	Data as of 1 January 1997: SNF mass, t	Data as of 1 January 1997: FP activity, EBq[a]
Balakovo			216	42.3
Beloyarsk			256	2.6
Bilibino			110	0.68
Kalinin			195	42.0
Kola			156	22.8
Kursk			2,867	57.0
Leningrad			3,662	73.0
NovoVoronezh			190	10.6
Smolensk			1,449	48.0
Armenia	77	0.7		
Zaporozhie	291	51.4		
Ignalina	637	38.5		
Rovno	1,52	16.4		
Khmelnitsk	89	10.3		
Chernobyl	2,081	39.6		
South Ukraine	173	20.9		

[a]EBq = 10^{18} Bq.

Table 4.3. Catcgorics of RW.

RW category (group)	Liquid: By volume activity (Bq/l)	Solid: By specific activity (Bq/kg)	Solid: By surface contamination (part/cm^2/min) β emitters[a]	Solid: By gamma-dose rate (mSv/hr)
Low active (low)	$< 3.7\times10^5$	From 7.4×10^4 to 3.7×10^6	From 5×10^2 to 1×10^4	From 1×10^{-3} to 0.3
Medium active (medium)	From 3.7×10^5 to 3.7×10^{10}	From 3.7×10^6 to 3.7×10^9	From 1×10^4 to 1×10^7	From 0.3 to 10
High active (high)	$> 3.7\times10^{10}$	$> 3.7\times10^9$	$> 10^7$	> 10

[a]α-emitter activity and contamination values are 10 times lower than the β-emitter activity and contamination values.

upon their specific activities, or gamma-ray dose rates formed at a distance of 0.1 m, or surface contamination by radionuclides. *Table 4.3* illustrates this classification.

At present, solid wastes at NPPs are divided into different categories only by measurement of gamma-ray dose rates at 0.1 m from the container, cask, package, bag, etc., which is the simplest method used for this purpose.

SRWs are formed during scheduled repairs and overhauls (process equipment, control and measurement detectors, pipes, tools, clothes, etc.) as well as in the process of decontamination of liquid and gaseous RWs (filters, sorbents, ion-exchange resins, etc.).

Liquid wastes are produced during NPP operations such as maintaining the water-chemical conditions of the process circuits of the reactor unit, and purification of waters with low salt concentrations during the decontamination of equipment, rooms, and clothes. By their physical and chemical composition, LRWs are homogeneous, aqueous, and organic solutions, as well as heterogeneous systems (pulps, emulsions, suspensions) in acidic and alkaline states. The nuclide composition of NPP RWs involves some FPs (cesium, strontium, and iodine) and products of activation of the reactor plant structural materials (cobalt, nickel, manganese). Among the radionuclides of this set, there are no significant amounts of isotopes with half-lives exceeding 30 years. This makes it possible to consider the possibility of burying the RWs of this category in the repositories and securing their isolation for about 300–500 years, as within this period they nearly completely decay.

The rate of RW production at NPPs depends on the type of the reactor unit, quality of NPP operation, and many other factors. The quantitative indices of RW formation per year and some of their characteristics for different reactor types are given in *Table 4.4*.

Table 4.5 presents data on SRWs accumulated in the storage facilities at the sites of the Russian and former Soviet NPPs. *Table 4.6* lists the general data on RWs produced in NPPs of the FSU and Russia.

At present, the storage facilities in most Russian NPPs are filled with liquid wastes to about 75–100% capacity, and with solid wastes to over 60% capacity. The amount of high-active wastes, by volume, is no more than 1%. Some numerical data in the tables are continuously changing because operating NPPs are producing RWs, whereas stored RW is subject to reprocessing.

Table 4.4. Quantitative indices of RW production rates at NPPs.

No.	RW type	VVER–440	VVER–1000	RBMK–1000
1.	Distillation residue transferred to repositories, m^3/yr	120–170	220–300	1,000–1,200
	Average salt content of stillage residue, g/l.	300–400	300–400	200–250
	Total salt quantity, t/yr	50	90	250
	Specific activity of the stillage residue, Bq/l.	1.85×10^6	1.85×10^6	1.85×10^6
2.	Low-active sorbents, m^3/yr	8.9	16	62
	Specific activity, Bq/kg	1.11	3.7×10^7	3.7×10^7
3.	High-active sorbents, m^3/yr	3.00	5.3	22
	Specific activity, Bq/kg	1.85×10^9	1.85×10^9	1.85×10^9
4.	Pearlite, m^3/yr	–	–	9.0
	Specific activity, Bq/kg	–	–	7.4×10^7
5.	SRW, m^3/yr	200	300	400

4.2.2 Gas-aerosol releases and liquid waste disposal from NPPs

Gas-Aerosol Releases from NPP

By convention, the sources of gas-aerosol releases from NPPs can be divided into two large groups: coolant leaks into NPP rooms and process blowouts from NPP water system tanks.

Coolant leaks are usually accompanied by radioactive aerosols of long-lived nuclides (LLNs) and radioactive iodine (I) releases. The process blowouts lead primarily to releases of noble radioactive gases (NRGs).

As these three constituents (LLN, I, and NRG) of releases to the atmosphere differ in physical states, they possess different physical and chemical properties and different methods are used to reduce their chances of escape to the atmosphere through the NPP vent stack. The vent air is cleansed of aerosols and iodine by installation of aerosol and special "iodine" filters into the exhaust ventilation systems. Process blowouts are cleaned from NRGs in special radiochromotographic systems. The exposure chambers are used only at first-generation power units (Kursk and Leningrad NPPs). The gas-aerosol release cleaning systems currently in operation at NPPs ensure 80–99.9% cleaning efficiency.

Table 4.5. SRW storage facilities.

NPP	Date of commis-sioning	Waste category	Total activity, TBq	RW characteristics				Rated capacity, m^3	Actual filling, m^3	Filling rate, m^3/yr	Protection barrier
				Type	Vol. activity, GBq/m^3	Chem. form.	Solubility				
Balakovo[a]	1985	Medium and low active	1.3	FP and corrosion products	0.2	Bound	Low	18,680	72,90	810	Concrete and wet seal
Beloyarsk[a]	1964	" "	6.3	" "	0.4	" "	" "	23,760	14,300	480	" "
Bilibino[a]	1974	" "	7.5	" "	3.5	" "	" "	3,150	2,150	100	" "
Kalinin[a]	1984	" "	1.7	" "	0.3	" "	" "	6,520	5,080	500	" "
Kola[a]	1973	" "	1.6	" "	0.2	" "	" "	19,900	7,040	335	" "
Kursk[a]	1976	" "	6.7	" "	0.3	" "	" "	21,760	21,640	1,200	" "
Leningrad[a]	1974	" "	18.5	" "	1.2	" "	" "	24,000	15,720	925	" "
NovoVoronezh[a]	1963	" "	6.7	" "	0.2	" "	" "	39,860	28,180	900	" "
Smolensk[a]	1982	" "	3.4	" "	0.3	" "	" "	15,150	10,630	760	" "
Total[a]			53.7					172,780	112,030		
Armenia[b]	1976	Low and medium active	0.7	" "	0.4	" "	" "	12,000	1,830	100	" "
Zaporozhie[b]	1984	" "	1.7	" "	0.4	" "	" "	19,500	4,250	700	" "
Ignalina[b]	1984	" "	1.9	" "	0.3	" "	" "	5,700	5,500	900	" "
Rovno[b]	1980	" "	1.3	" "	0.4	" "	" "	5,700	3,500	300	" "
Khmelnitsk[b]	1987	" "	0.4	" "	0.4	" "	" "	4,000	980	250	" "
Chernobyl[b]	1977	" "	5.2	" "	0.4	" "	" "	13,890	13,890	–	" "
South Ukraine[b]	1982	" "	2.3	" "	0.4	" "	" "	12,000	5,700	600	" "
Total[b]			13.5					72,790	35,650		

[a] As of 1 January 1995.
[b] As of 1 January 1991.

Table 4.6. LRW storage facilities.

NPP	Date of commissioning	Source	Total activity, TBq	RW category	Actual filling, m^3	RW characteristics: Type	Vol. activity, GBq/m^3	Chem. form	Fire risk	Explosion risk
Balakovo[a]	1985	Distillation residue	3.3	Medium- and low-active	1,650	FP and corrosion products	2.0	Ion precipitation	No	No
		Ion-exchange resins (IEXR)						IEXR-bound	No	No
Beloyarsk[a]	1964	" "	51.8	" "	5,450	" "	9.5	" "	" "	" "
Bilibino[a]	1974	" "	3.3	" "	690	" "	4.8	" "	" "	" "
Kalinin[a]	1984	" "	29.6	" "	2,780	" "	10.6	" "	" "	" "
Kola[a]	1973	" "	70.3	" "	5,240	" "	13.4	" "	" "	" "
Kursk[a]	1976	" "	45.0	" "	35,240	" "	1.3	" "	" "	" "
Leningrad[a]	1974	" "	18.5	" "	11,920	" "	1.6	" "	" "	" "
NovoVoronezh[a]	1963	" "	81.4	" "	6,620	" "	12.3	" "	" "	" "
Smolensk[a]	1982	" "	150.2	" "	11,730	" "	12.8	" "	" "	" "
Total[a]			453.4		81,320					
Armenia[b]	1976	" "	37.0	" "	3,530	" "	10.5	" "	" "	" "
Zaporozhie[b]	1984	" "	14.8	" "	1,300	" "	11.4	" "	" "	" "
Ignalina[b]	1984	" "	44.4	" "	3,960	" "	11.2	" "	" "	" "
Rovno[b]	1980	" "	48.1	" "	4,500	" "	10.7	" "	" "	" "
Khmelnitsk[b]	1987	" "	7.4	" "	600	" "	12.3	" "	" "	" "
Chernobyl[b]	1977	" "	364.5	" "	40,400	" "	9.0	" "	" "	" "
South Ukraine[b]	1982	" "	40.7	" "	4,000	" "	10.2	" "	" "	" "
Total[b]			556.9		58,290					

[a] As of 1 January 1995.
[b] As of 1 January 1991.

Table 4.7. Admissible releases of NPPs currently in operation.

Standardized release	Admissible values
NRG	18.60 TBq/day
LLN	0.55 GBq/day
^{131}I	0.37 GBq/day

To reduce the radiation effect of the NPP on the population and the environment, the Soviet nuclear safety regulatory bodies established that, during normal operations, the NPP shall not cause any additional irradiation of the population with a dose exceeding 0.2 mSv/yr from gas-aerosol releases or a dose of 0.05 mSv/yr because of radionuclides contained in the effluents.

Russian nuclear safety regulatory bodies have established annual gas-aerosol release and liquid waste discharge limits, called admissible release (AR) and admissible discharge (AD) limits. However, up to now, no standard method for AR calculation has been adopted in Russia, and all NPPs use lower AR values that were established by the nuclear safety regulatory bodies in 1979. These values were adopted based on the reference values of actual releases from NPPs. The values of admissible releases per NPP power unit currently in operation are listed in *Table 4.7.*

The data on the actual values of daily average gas-aerosol releases from Soviet NPPs in 1991 and Russian NPPs in 1995 by standardized components (NRG, LLN, and I) are displayed in *Table 4.8.*

In 1994, a significant reduction (by about four times) of NRG releases to the atmosphere took place at Russian NPPs due to implementation of a system to suppress (by increasing the exposure before the discharge) Ar-41 activities, whose contribution to releases had exceeded 60%.

The calculation of the additional radiation dose to the population due to gas-aerosol releases in 1995 (listed in *Table 4.8*) gives a value of less than 0.01 mSv, which cannot be measured by any instrument at the level of the natural irradiation dose (about 1 mSv). These doses did not cause a change in the radiation situation in the areas near NPPs.

Liquid Waste Discharge from NPPs

For all Russian NPPs, nuclear safety regulatory bodies have established admissible RW discharges that take into account the specific features of water reservoirs where the wastewaters from each NPP are discharged.

Table 4.8. Daily average gas-aerosol releases from NPPs.

NPP,	NRG		LLN		^{131}I	
Number of units	TBq/day	% AR	MBq/day	% AR	MBq/day	% AR
Balakovo,[a] 3 units	0.03	0.04	0.12	0.01	0.47	0.03
Beloyarsk,[a] (1+2) units	0.01	0.04	–	–	–	–
Bilibino,[a] 4 units	0.80	4.35	–	–	–	–
Kalinin,[a] 2 units	0.06	0.15	0.16	0.01	4.07	0.55
Kola,[a] 4 units	0.07	0.18	0.56	0.05	3.31	0.45
Kursk,[a] 4 units	3.0	4.11	12.58	0.56	20.72	1.38
Leningrad,[a] 4 units	4.2	4.03	43.29	1.96	53.65	3.63
NovoVoronezh,[a] 5 units	0.07	0.18	3.26	0.29	0.71	0.10
Smolensk,[a] 3 units	2.79	5.02	3.22	0.19	20.39	1.84
Armenia,[b] 2 units	Units have been shut down.					
Zaporozhie,[b] 5 units	0.40	0.43	0.48	0.02	0.33	0.02
Ignalina,[b] 2 units	9.7	17.0	18.0	1.1	2.2	0.20
Rovno,[b] 3 units	0.20	0.51	0.49	0.04	1.1	0.14
Khmelnitsk,[b] 1 unit	0.09	0.49	0.42	0.08	0.36	0.10
Chernobyl,[b] 3 units	10.0	19.0	110.0	6.5	18.0	1.6
South Ukraine,[b] 3 units	0.20	0.30	0.24	0.02	0.03	0.003

[a] As of 1995.
[b] As of 1991.

In all possible kinds of water reservoir management, the addition to the background dose must not exceed 0.05 mSv/yr.

In the ideal case, the NPP process cycle in the primary and secondary circuits are closed, without production of wastewaters (i.e., the difference in the quantities of the water from the process circuits transferred for cleaning and that arriving at these circuits). However, because of different design solutions adopted at different power units and the levels of their operation, excess wastewater is produced. Having passed through cleaning in the evaporator and ion-exchange systems, this water accumulates in special (control) tanks and, after the radiation control, can be discharged from the NPP to surface water reservoirs.

For each discharge of wastewater, there must be a certificate on record specifying the volume of water to be discharged and the volumetric and total activity of all radionuclides in it.

The disposal of the NPP wastewater is permitted if two conditions are fulfilled:

Table 4.9. Discharge of radionuclides into the environment from NPP liquid waste disposals.

NPP	Volume of wastewater, thousands of m^3	Activity as percentage of AD
Balakovo [a]	7.4	13.2
Beloyarsk[a]	36.0	16.0
Bilibino[a]	1.9	0.03[c]
Kalinin[a]	73.0	45.7
Kola[a]	12.0	3.3
Kursk[a]	48.0	[d]
Leningrad[a]	5.0	0.2
NovoVoronezh[a]	61.0	8.0
Smolensk[a]	94.0	32.7
Armenia[e]	–	–
Zaporozhie[b]	22.0	7.2
Ignalina[b]	20.0	1.6
Rovno[b]	33.0	10.0
Khmelnitsk[b]	38.0	7.4
South Ukraine[b]	25.0	13.0

[a]As of 1 January 1995.
[b]As of 1 January 1991.
[c]By the main dose-producing radionuclide, ^{60}Co.
[d]Lower than the minimum detected activity.
[e]Units had been shut down and have since reopened.

- The volumetric activity of the water does not exceed the value established in the Standards of Radiation Safety (i.e., it does not fall into the category of RWs).
- The total activity of all nuclides discharged with the wastewater from the NPP from the beginning of the year does not exceed the AD.

The volume of wastewaters discharged from NPPs and the activity as a percentage of the AD value are given in *Table 4.9*.

It should be pointed out that the high percentages of discharges from some NPPs are due not to high values of the activity released (in absolute units they are about 1 GBq/yr), but rather to the strict standards currently in force in Russia.

4.2.3 Radiation incidents and accidents at NPPs

The International Nuclear Events Scale (INES) worked out by the International Atomic Energy Agency (IAEA) was adopted and has been used in Russia since 1991. Before then, a three-level scale of estimation of the magnitude of radiation accidents at NPPs had been used. According to the extent of which the quantity of RM released exceeded the designed boundaries, all accidents were classified as localized, local, or general.

- A localized accident is a failure in the NPP operation resulting in the release of radioactive products or ionizing radiation beyond the specified boundaries for equipment, process systems, buildings, and structures in amounts exceeding the standards established for normal operation.
- A local accident is a failure in the NPP operation that results in radioactive products escaping into the NPP sanitary-protection zone (SPZ) in amounts exceeding the standards established for normal operation.
- A general accident is a failure in the NPP operation that results in radioactive products escaping beyond the boundaries of the NPP SPZ in amounts exceeding the standards established for normal operation.

The relation between the previously used NPP accident scale and the current INES can be tentatively equated as follows:

Soviet (Russian) scale before 1991	INES
Localized accident	Level 0, 1, 2 (incident)
Local accident	Level 1, 2, 3 (incident)
General accident	Level 3–7 (accident)

When designing an NPP, a set of initial events (equipment failures, personnel errors, etc.) leading to accidents is analyzed to substantiate the NPP's safety, and safety systems (control, protection, localizing, etc.) are designed to prevent the development of accidents and to mitigate their consequences. The requirements for the safety of the NPP reactor units were continuously increased, which was reflected in normative documents (NDs) on NPP safety and introduced to the designs. With regard to the safety level, all Soviet-designed NPP units can be divided into two generations: first-generation power units developed and put into operation before the main nuclear safety NDs were issued (Units 1 and 2 of Beloyarsk, and NovoVoronezh, Kola, Leningrad, and Kursk NPPs); and second-generation power units (all later ones) designed and constructed in accordance with NDs corresponding to the main international approaches, such as OPB–72, OPB–82, OPB–88, and others.

Safety enhancements for all operating NPPs were largely completed before the disintegration of the Soviet Union; these enhancements incorporated data from the Chernobyl NPP accident and additional safety analysis. The higher nuclear safety level of the NPPs with RBMK reactors was reached due to the reduction in the positive void coefficient of reactivity, an increase in the response of the regular system of emergency protection, and the introduction of an additional quick-acting emergency protection. As a result, the possibility of an accident occurring due to a rapid uncontrolled transient in the reactor, similar to the Chernobyl accident, has been greatly reduced. At NPPs with VVER–1000 reactors, measures have been introduced ensuring the required reliability of the mechanical emergency protection. Some longer-term measures for the enhancement of the safety and reliability of the NPP units with both RBMK and VVER reactors have been worked out and are now being realized. These measures include strengthening VVER reactor vessels to make them more resistant to fractures, and the refit of Units 1 and 2 of the Leningrad and Kursk NPPs.

The supervisory body for nuclear safety regulation (RF Gosatomnadzor) has established a special operation regime for NPPs with first-generation power units, including issuing annual reports containing analysis of the actual state of safety and obtaining permits for further operation. Similar practices relative to NPPs that do not completely meet the current safety requirements also exist in other countries such as Great Britain, France, and the USA. These countries were the first to develop nuclear power, and some NPPs are still equipped with out-of-date power units.

The information on events (incident and accidents) at the Soviet/Russian NPPs that have led to radiation consequences is given in *Table 4.10*. With the exception of the Chernobyl accident, none of the accidents listed in *Table 4.10* resulted in changes in the radiation situation outside the NPP site, and no special measures were required for protecting the general population. The Chernobyl accident, whose initial events,

Table 4.10. Incidents and accidents at NPPs in the Soviet Union/Russia.

No.	NPP, unit number	Date	Initial event	Consequence scale
1.	Leningrad, 1	30.11.75	Rupture of process channel (RPC)	Local accident
2.	Rovno, 1	22.01.82	Detachment of collector covers of SG 1, 3–6	Local accident
3.	Chernobyl, 1	09.09.82	RPC	Local accident
4.	Chernobyl, 1	25.02.83	RPC	Local accident
5.	Chernobyl, 4	26.04.86	Thermal explosion of the reactor	General accident
6.	Ignalia, 2	1988	Fire damages 650 control cables interrupting forced cooling and disabling safety monitor systems	Local accident
7.	Beloyarsk, 2	1989	Fuel channel ruptures with fuel failure	Local accident
8.	Kalinin, 1	06.01.90	SG–4 overfills with operation of quick-acting pressure reducing plant (BRU–A)	Local accident
9.	Ignalina, 2	05.06.91	Irradiation of three men with dose exceeding 5cSv	Local accident
10.	Leningrad, 3	24.03.92	RPC	INES Level 3
11.	Kalinin, 2	02.07.92	Leak of radioactive water from bituminous RW storage to NPP site	INES Level 1
12.	Beloyarsk, 1, 2	23.12.92	LRW storage overfill and release of radionuclides to the ground	INES Level 2
13.	Beloyarsk, 3	07.10.93	Depressurization of pipe of primary circuit sodium decontamination	INES Level 1
14.	Kola, 2	02.03.94	Depressurization of primary circuit make-up pipe in power unit shutdown	INES Level 2
15.	Kursk, 4	21.11.95	Irradiation of two men with dose exceeding 5cSv	INES Level 2

Table 4.11. Average characteristics of radiation exposure of liquidators.

Year of arrival	N (prs)	D (cGy)	σ (cGy)	T (days)	σ (days)	P (mR/hr)	σ (mR/hr)
1986	46,575	15.9	8.3	7	72	19.2	45.5
1987	48,077	9.0	5.8	79	59	6.8	19.8
1988	18,208	3.3	3.5	106	58	2.0	14.8
1989	5,475	3.2	3.6	102	60	1.9	6.7
1990	1,004	3.7	2.9	86	58	2.5	7.3

Abbreviations: N = number of liquidators, persons; D = average absorbed dose, cGy; T = average time of residence in the contaminated zone, days; P = average EEDR, mR/hr; σ = D,T,P distributions standard deviation.

development, and consequences have been studied in the USSR/Russia and in many other countries, has been described in detail in the literature.

This analytical review is confined to presenting new data on the radiation doses received by people who participated in the operations to mitigate the consequences of the accident, and describing the impact of the accident on the environment.

Radiation Doses from the Chernobyl Accident

In 1986, after the Chernobyl accident, the All-Union Distributed Registry (AUDR) of individuals who were subjected to irradiation was set up in the Soviet Union. By the time of the disintegration of the Soviet Union, the AUDR database had medical and dosimetrical information on 660,000 persons, including 275,000 participants in the mitigation of the Chernobyl NPP consequences. The Russian State Medical and Dosimetric Registry (RSMDR), which was established in 1992, includes 370,000 Russian citizens who were subjected to irradiation in nuclear accidents. As of July 1995, the RSMDR contained the data on the health of 159,027 emergency workers. For 125,771 emergency workers, the external radiation dose values are put into the RSMDR based on the health certificates delivered in the zone of the mitigation of accident consequences (MAC) operations.

To obtain a more detailed analysis of radiation dose rates, efforts were made to establish the geographical coordinates of the settlements where the emergency workers lived. Such data were gathered on 119,782 emergency workers with the available dosimetric data. For further analysis, records were chosen from the RSMDR database with the external radiation dose not exceeding 50 cGy. The radiation doses of 366 emergency

Table 4.12. Average characteristics of radiation exposure of liquidators for different distances from the Chernobyl NPP, 1986–1988.[a]

R_1–R_2 (km)	0–10	10–30	30–50	50–100	100–250
1986					
N (prs)	2,461	20,362	16,368	6,976	390
D (cGy)	14.8	14.8	17.8	15.4	8.5
σ (cGy)	9.4	8.8	7.2	7.4	9.0
P (mR/hr)	33.6	21.0	16.4	16.0	12.0
σ (mR/hr)	84.5	53.0	24.9	38.8	27.7
1987					
N (prs)	1,612	15,004	18,417	12,800	244
D (cGy)	7.5	8.5	9.1	9.6	5.2
σ (cGy)	6.5	6.1	5.5	5.5	5.7
P (mR/hr)	5.7	6.8	7.0	6.8	2.3
σ (mR/hr)	12.2	19.9	21.9	17.0	8.5
1988					
N (prs)	336	5,646	6,325	5,658	243
D (cGy)	4.2	3.9	3.3	2.7	1.9
σ (cGy)	4.5	3.9	3.3	3.1	3.3
P (mR/hr)	4.0	2.4	1.8	1.8	1.0
σ (mR/hr)	16.4	8.4	4.9	24.3	3.9

[a] See *Table 4.11* for abbreviations.

workers contained in the RSMDR were higher than 50 cGy. (Unless these cases are not specified, they are not included in *Table 4.11*.)

Table 4.11 lists the average values and standard deviations of the radiation doses for emergency workers, time of their residence in the zone where the work was carried out, and effective exposure dose rate (EEDR) depending on the date of their arrival at the contaminated zone. From the data of *Table 4.11*, conclusions can be made about the dynamics of the average radiation dose rates in performing work in the accident zone. Some increase in the average dose for emergency workers arriving at the zone in 1990 can be accounted for by the fact that the bulk of the work in the zone had been completed in the main and, therefore, only professionals were sent to the zone for special operations.

Table 4.12 displays similar average characteristics for different distances from residence or working area to the Chernobyl NPP from 1986 to 1988. (R_1–R_2 is the distance from the residence or work area to the

Chernobyl NPP in kilometers.) From the data in *Table 4.11*, the collective radiation dose received by most MAC participants can be determined:

1986: 7,405.4 person-Sv.
1987: 4,327.0 person-Sv.
1988: 600.9 person-Sv.
1989: 175.2 person-Sv.
1990: 371.5 person-Sv.
Total from 1986 to 1990: 12,880 person-Sv.

Further work on the systematization of the data on the dose rates received from the Chernobyl accident has been conducted at the Russian Academy Medical Science (RAMS) Medical Radiological Research Center in Obninsk.

Territories Contaminated with Radionuclides from NPP Accidents

The accident at Unit–4 of the Chernobyl NPP began with a violent thermal explosion in the core. A high-temperature fireball containing a complete set of radionuclides that had accumulated in the reactor was discharged into the atmosphere. A large radioactive cloud formed after the explosion and began to spread in the prevailing wind direction, which initiated radioactive contamination of the environment. The fire on the graphite stack following the explosion was maintained by the high energy released from the continuing decay of radionuclides emerging from the reactor. This resulted in a radioactivity discharge into the atmosphere in the form of a jet that continuously changed its direction, depending on wind direction. The intense jet discharge was observed for about 10 days.

On 26 April 1986, a part of the air mass trajectories was directed from the Chernobyl NPP through Belorussia and the Baltic states to Finland, Sweden, and Norway (most likely, it was an explosion discharge). Another portion of the trajectories went through the Ukraine and Poland to Germany, Denmark, Belgium, the Netherlands, and northern France. On 27 April, the atmospheric trajectories went again from the Chernobyl NPP through Belorussia and Scandinavian countries, but then turned toward the Kola Peninsula to the east. On 28 April, the situation was nearly calm with the wind still blowing to the east; on 29 April, the trajectories were mainly toward the east; and on 30 April, they turned to the south and then to the southwest and west. Consequently, the contaminated air had spread over the European part of the Soviet Union and over all of Europe in the first days of the accident.

The evaluation, measurement, and analysis of the radiation situation and radioactive contamination of the environment began on day one of the Chernobyl accident and continue today. Prompt radiometric (and later, gamma-spectrometric) surveys were carried out using ground and aviation methods. The first complete map of the near-ground trace (at up to 100 km from the accident) was made on 1 May, and submitted to the Governmental Commission on 2 May 1986. In the first days, such maps were made practically every day. In May 1986, the Governmental Commission evacuated the population from the zone where the annual dose might be 10 rem or higher (an evacuation criterion established by the Soviet Ministry of Public Health). This dose corresponded to earth surface radiation levels (at a height of 1 m) of 5 mR/hr as of 10 May 1986. The Governmental Commission established alienation zones for areas with radiation levels greater than 20 mR/hr as of 10 May 1986, and partial evacuation zones for areas with radiation levels from 3–5 mR/hr as of the same date. On 10 May 1986, the alienation zones had an area of 1,100 km^2, and evacuation zones had an area of 2,940 km^2. At the same time, the Soviet territory contaminated with radioactive discharges at levels exceeding 0.2 mR/hr from the Chernobyl NPP was about 50,000 km^2.

As early as the beginning of May, full-scale studies of radionuclide contamination of the environment began (although some samples had been analyzed from the first days of the accident). Many radionuclides (^{89}Sr, ^{90}Sr, ^{91}Y, ^{95}Zr, ^{95}Nb, ^{99}Mo, ^{103}Ru, ^{106}Ru, ^{125}Sb, ^{131}I, ^{132}Te, ^{132}I, ^{133}I, ^{134}Cs, ^{136}Cs, ^{137}Cs, ^{140}Ba, ^{140}La, ^{141}Ce, ^{144}Ce, ^{239}Np, and some others) were measured in the atmosphere. The studies revealed that relative to ^{95}Zn and ^{144}Ce, ruthenium isotopes exhibited an increased volatility, and ^{131}I, ^{132}Te, ^{132}I, ^{133}I, ^{134}Cs, ^{136}Cs, and ^{137}Cs exhibited a high volatility. Measurements of concentrations of ^{239}Pu and ^{240}Pu in the near-earth surface air level beyond the 30-km zone showed that at all sampling points their concentrations in the air were lower than permissible concentrations (1.11×10^{-8} Bq/l), even in the first month after the accident. In soil samples taken within several days of the accident, all of the above radionuclides as well as ^{238}Pu and ^{239,240}Pu isotopes were detected. Somewhat later ^{241}Pu, ^{241}Am, ^{242}Cm, and ^{244}Cm were identified.

A noticeable fractionation of radionuclides, differing depending on the radioactivity transfer direction (south, west, north), was detected on

Table 4.13. Relative radionuclide composition of radioactive fallout in May–June 1986 in the zone nearest to the Chernobyl NPP.

Radio-nuclide	Half-life	Southern sector		Western sector		Northern sector	
		$(A_i/A_{95})_E$	$f_{i,95}$	$(A_i/A_{95})_E$	$f_{i,95}$	$(A_i/A_{95})_E$	$f_{i,95}$
^{89}Sr	51 days	1.6	3.0	0.6	1.3	0.3	0.6
^{90}Sr	28.5 year	0.16	2.0	0.03	0.4	0.03	0.4
^{91}Y	58.5 day	n.a.	n.a.	0.6	0.9	0.5	0.8
^{95}Zr	65 days	1.0	1.0	1.0	1.0	1.0	1.0
^{99}Mo	2.73 days	n.a.	n.a.	1.4	1.6	n.a.	n.a.
^{103}Ru	39 days	0.7	0.7	0.7	0.7	1.6	1.5
^{106}Ru	368 days	0.2	0.35	0.3	0.5	0.4	0.7
^{131}I	8.01 days	0.6	1.0	0.6	0.6	6.0	10
^{132}Te	3.27 days	0.9	1.2	3.8	5.3	10	14
^{134}Cs	2.06 year	0.015	0.1	0.14	0.9	0.18	1.2
^{137}Cs	30.1 year	0.04	0.4	0.2	1.8	0.3	2.7
^{140}Ba	12.6 days	0.9	1.0	1.6	1.5	1.1	1.0
^{141}Ce	32.5 days	1.1	1.2	1.1	1.2	1.1	1.2
^{144}Ce	284 days	0.6	0.6	0.7	0.7	0.6	0.6
^{239}Np	2.35 days	n.a.	n.a.	9.0	n.a.	n.a.	n.a.

n.a. = not available.

the earth surface even in May 1986. The coefficient of fractionation of an i-th radionuclide relative to ^{95}Zr may be determined as

$$f_{95}^{i} = \frac{(A_i/A_{95})_E}{(A_i/A_{95})_T} ,$$

where A_i and A_{95} are activities of the i-th radionuclide and ^{95}Zr, correspondingly, at the same moment in time; $(A_i/A_{95})_E$ is the ratio of activities determined experimentally at a certain point in the environment (for instance, in a soil specimen); and $(A_i/A_{95})_T$ is the ratio related to the total amounts of both radionuclides, formed as a result of an explosion (determined by calculation).

The coefficient of fractionation shows in which way the initial ratio $(A_i/A_{95})_T$ changes within the environment, depending on various processes of the particles' formation and their spread in the environment.

The radionuclide composition of contamination in the zone nearest to the Chernobyl NPP is shown in *Table 4.13*, which lists the data on all the samples for which the gamma-ray dose rate measurements at a height of 1 m were made in May–June 1986. Doses ranged from 2 mR/hr to

10 mR/hr, although there were doses of higher and lower radiation levels. As shown in the table, the radionuclide composition of fallout in the nearest Chernobyl NPP zone is represented by a complete set of fission fragments (additionally, the quantity of ^{134}Cs and ^{239}Np do not differ much from the quantities in which they had accumulated in the reactor at the onset of the accident). Only ^{131}I and ^{132}Te in the north sector are an exception.

By the 15th day, the fraction of volatile nuclides in the nearest section of the trace (up to 40 km) was up to 5% of ^{132}Te, 5.1% of ^{131}I, and 1.6% on average. This last value is estimated by the relation of the total energy releases of radionuclides on the trace and in the reactor (at the moment of onset of the accident) reduced to the 15th day.

From 5 May 1986, data showing unusual radionuclide compositions of the air contamination began to be reported from research ships that were carrying out work in the North Atlantic and the Pacific. According to the data from the ships, only nine Chernobyl radionuclides were detected: ^{89}Sr, ^{91}Y, ^{103}Ru, ^{131}I, ^{132}Te, ^{140}Ba, ^{134}Cs, ^{136}Cs, and ^{137}Cs, with cesium isotopes predominant. Thus, it was found that the largest fraction of radioactive particles transferred with air currents for long distances were volatile radionuclides: krypton and xenon isotopes, ^{131}I, ^{132}Te, ^{132}I, and ^{133}I. Such isotopes as ^{134}Cs, ^{136}Cs, and ^{137}Cs, as well as ^{103}Ru and ^{106}Ru, represented a somewhat smaller fraction. ^{91}Y and strontium isotopes ^{89}Sr and ^{90}Sr were much less volatile; they were recorded in trace quantities.

In May 1986, the criteria for the maximum permissible contamination of the environment with long-lived radionuclides were established (based on the standards of maximum permissible concentrations for ^{239,240}Pu in air and maximum permissible dose rates). They were 2.59×10^{11}–5.55×10^{11} Bq/km^2 for ^{137}Cs, 1.11×10^{11} Bq/km^2 for ^{90}Sr, and 3.7×10^{9} Bq/km^2 for ^{239,240}Pu. In terms of the criteria established, plutonium radionuclides, ^{238}Pu, and ^{239,240}Pu (3.7×10^{9} Bq/km^2) were found to be mainly localized within the 30-km zone of the Chernobyl NPP, and ^{90}Sr (1.11×10^{11} Bq/km^2) was within about 60 km. Studies carried out on the Russian territory revealed that in none of the most intense spots of radioactive contamination did the radioactive contamination exceed 1.11×10^{11} Bq/km^2 for ^{90}Sr or 3.7×10^{9} Bq/km^2 for ^{239,240}Pu.

In the initial period of the Chernobyl accident, the radioactive contamination of the territory was essentially due to the release of short-lived radionuclides that later lost their ecological significance. In a year after the accident, the radiation situation on the largest part of the European territory of the Soviet Union was determined by long-lived isotopes of cesium, specifically ^{137}Cs and ^{134}Cs; in three years it was determined by ^{137}Cs. Therefore much attention was given to map-making of environmental contamination with just ^{137}Cs. Special state financing was provided for the investigations and for making state maps of environmental contamination. These maps required a wide scientific generalization of the information obtained, using unified methodology from different activities and different methods that use up-to-date equipment.

In accordance with the Russian Federation "Law On Special Protection of Persons Exposed to Radiation from the Catastrophe at Chernobyl NPP," the isoline 1 Ci/km^2 (37 kBq/m^2) for ^{137}Cs is the lower threshold for establishing privileges for the population. The law specifies the dwelling zone with a privileged socioeconomic status (1–5 Ci/km^2 or 37–185 kBq/m^2); the dwelling zone with the right of resettlement (5–15 Ci/km^2 or 185–555 kBq/m^2); and the zone of mandatory evacuation (more than 15 Ci/km^2 or more than 555 kBq/m^2). The annual dose of internal and external irradiation of the population must not exceed 0.1 rem. Similar laws have been adopted in Ukraine and Belorussia.

A great amount of work was carried out to obtain information on the contamination levels in settlements. In Russia, 11,457 settlements in 23 administrative regions were inspected in 1995. For the whole post-accident period, about 90,000 soil samples were taken that were subjected to the gamma-spectrometric, and partly to radiochemical, analysis. Based on these data, levels exceeding 1 Ci/km^2 or 37 kBq/m^2 were observed in 4,581 settlements. The calculations revealed that an excess of the permissible irradiation level of 0.1 rem can be expected in 527 settlements in 1995. The contribution of ^{239}Pu and ^{240}Pu is less than 1%.

A number of investigations on the contamination of large and small territories of the European part of the Soviet Union, and later in Russia, were accomplished using the aero-gamma-spectrometric method. These investigations permitted the data on the settlement contamination to be supplemented with the data obtained outside the settlement territories. The aero-gamma-spectrometric surveys of environmental contamination with various radionuclides (^{95}Zr, ^{95}Nb, ^{140}La, ^{103}Ru, ^{134}Cs, ^{137}Cs, and ^{144}Ce) have been carried out since late May 1986. As of early 1997, the

European part of the FSU and Siberian territory to about the geographical latitude of the Yenisei have been covered by the surveys.

The aero-gamma-spectrometric survey and soil sampling were performed in the administrative regions. The interroute distances in the aero-gamma-spectral survey were determined by the levels of expected radioactive contamination: greater than 15 Ci/km^2 for ^{137}Cs, 0.5 km (surveying scale 1:50,000); 0.5–15 Ci/km^2, 2 km (1:200,000); less than 0.5 Ci/km^2, 10 km (1:1,000,000). The elementary route sections where the gamma-ray spectra are recorded during the flight are 0.1 km, 0.4 km, and 2.0 km, respectively. To construct a map of the European part of the FSU (see *Color Plate 1*), 800,000 complete spectra of aero-gamma-spectral measurements (440,000 spectra in the European part of Russia) were obtained and used. A special ground test accompanying the aero-gamma-spectral survey was carried out at several thousand points. In addition, the results of the analysis of tens of thousands of soil samples taken in settlements were used to construct the maps. The relative mean-square deviations in the ^{137}Cs contamination values taken from the aero-gamma-spectral and ground data did not normally exceed 40%; as a rule, they were between 20% and 25%. The systematic discrepancies were taken into account in constructing the maps.

Color Plate 1 presents the summary map of the ^{137}Cs contamination of the European part of the FSU as of 1992. The range of mapped contamination levels varies from a background level lower than 0.1 Ci/km^2 resulting from the atmospheric nuclear tests, mainly in the 1960s and 1970s, to levels reaching 500 Ci/km^2 and higher in the zones nearest Chernobyl NPP.

On the territories with contamination levels exceeding 0.2 Ci/km^2, fallout from the Chernobyl NPP accident is continuously present. This contamination was recorded on a 1,011,400-km^2 area, which constitutes about 23% of the European part of the FSU. The territories of Russia, Ukraine, and Belorussia with contamination levels above 1 Ci/km^2 constitute 145,300 km^2 – that is, about 3.3% – of the European territory of the FSU.

The summary data on the distribution of the ^{137}Cs contaminated areas in the whole European part of the FSU (1993) are listed in *Table 4.14*. The percentage of ^{137}Cs discharged from the reactor that fell on the European territory constituted about 15% of the amount having accumulated

Table 4.14. ^{137}Cs contaminated areas of the European part of the FSU as of 1993, thousand km^2.[a]

	Contamination level, Ci/km^2						
Region	0.1–0.2	0.2–0.5	0.5–1	1–5	5–15	15–40	>40
European Russia	877.5	365.3	112.80	48.80	5.72	2.1	0.3
Ukraine	142.8	186.8	72.50	37.20	3.20	0.9	0.6
Belorussia	14.8	67.2	26.40	29.90	10.20	4.2	2.2
Moldavia	10.4	23.0	0.15	0.06	n.a.	n.a.	n.a.
Baltic Republics	81.6	11.9	n.a.	n.a.	n.a.	n.a.	n.a.
Total	1,127.1	654.2	211.85	115.96	19.10	7.2	3.1

[a]The total area of the European part of the FSU is 4.45 million km^2.
n.a. = not available.

in the reactor by the time of the accident, or about 7×10^{16} Bq. Nearly 4×10^{16} Bq of this amount was deposited on Soviet territory.

After the accident at the Chernobyl NPP, radiological inspections of forests and arable lands were carried out. The results of these inspections as of 1993 are listed in *Tables 4.15* and *4.16*.

The radioactive contamination of water reservoirs, including rivers, on the territories of the Russian Federation, Ukraine, and Belorussia has been continuously monitored since the beginning of the accident, particularly where cesium hotspots in the water catchment areas have been observed.

The Chernobyl NPP is located on a territory with a developed hydrographic network. The 30-km zone and the main part of the radioactive trace are in the catchment basin of the Dnieper and its tributaries. To control the situation, a monitoring system was quickly deployed in the post-accident period, which covered all large and small water courses in the zone of radioactive contamination, all water storages of the Dnieper cascade, and the Baltic, Black, and Azov Seas.

Table 4.17 lists the maximum concentrations in the rivers nearest to the Chernobyl NPP during the observation periods beginning on 26 April 1986.

The total activity of the amount of radionuclides falling into the water from the clouds and jet stream was 7.4 PBq – the proportion of ^{90}Sr being about 0.037 PBq. The concentration of radionuclides in water reservoirs is determined by wash-off of about 0.5% of all ^{90}Sr and 0.1% of ^{137}Cs from the water catchment area during the first few post-accident years.

Table 4.15. ^{137}Cs contaminated arable lands, thousand ha, Ci/km^2.

	1–5	5–15	15–40	>40
Bryansk region	401	185	98	17
Kaluga region	128	33	1	–
Tula region	653	126	–	–
Orlov region	396	23	–	–

Table 4.16. ^{137}Cs contaminated forests, thousand ha, Ci/km^2.

	1–5	5–15	15–40	>40
Area	871.2	85.6	42.0	2.5

Table 4.17. Maximum radionuclide concentrations in the rivers of the vicinity of Chernobyl, Bq/l.[a]

River	Date	^{131}I	^{140}Ba	^{95}Zr	^{103}Ru	^{141}Ce
Pripyat	02.05.86	4,440	2,220	1,550	810	n.a.
Teterev	03.05.86	2,000	1,260	1,440	703	666
Irpen	06.05.86	1,850	1,110	810	n.a.	n.a.
Desna	26.04.86	666	520	370	122	337
Dnieper (Obolon)	03.05.86	1,300	703	1,410	n.a.	333

[a]The values of concentrations of medium- and long-lived radionuclides such as ^{144}Ce, ^{106}Ru, ^{90}Sr, ^{134}Cs, and ^{137}Cs are not presented in the table because from 26 April to 6 May 1986 their contribution to human exposure was insignificant.

This wash-off could not form water contamination that would exceed established standards.

The isotopic analyses of seawater samples, suspended solid materials, and sea-floor deposits revealed that the concentrations of radioisotopes of Chernobyl origin, such as $^{141,144}Ce$, $^{103,106}Ru$, ^{95}Zr, ^{95}Ni, $^{134,137}Cs$, ^{90}Sr, $^{238,239,240}Pu$, and tritium, are 100 times lower than the maximum permissible levels for drinking water and, therefore, are not harmful to human health.

The evaluation of the contamination of the bottom deposits of Dnieper's reservoirs was made in May 1986. It revealed that the bottom grounds (silt) in the Kiev water storage facility in the section adjacent to the mouth of the Pripyat are the most contaminated. In the southern part of the Kiev storage facility, as well as in the Kanev storage facility, the contamination level was found to be lower by tens and even hundreds of times.

Table 4.18. Estimated collective dose commitments for the most contaminated regions of Russia, during the 50-year period after the Chernobyl NPP accident.

Region	Population, 1,000 persons	Maximum external exposure, cSv	Average external exposure, cSv	Collective external exposure, 1,000 person-Sv
Bryansk	515	40	3.60	20.00
Tula	700	15	1.70	12.00
Oryol	352	15	1.05	3.70
Kaluga	168	12	1.50	2.50
Ryazan	203	10	1.40	2.80
Belgorod	102	7	1.25	1.30
Lipetsk	96	6	1.25	1.20
Kursk	61	5	1.25	0.75
Penza	45	5	1.25	0.55
Tambov	24	5	1.25	0.30
Voronezh	37	5	1.20	0.45
Leningrad	15	5	1.35	0.20
Ulyanovsk	7	5	1.45	0.10
Smolensk	5	5	1.50	0.75
Mordovian Republic	20	5	0.50	0.10
Total	2,350		2.00	47.15

After the accident, dozens of observation wells were drilled several hundred meters from one another to the depth of the water-bearing level around the Chernobyl NPP to monitor water quality. The observation of ^{90}Sr concentrations in the water for several post-accident years showed that these concentrations do not exceed background values.

Based on the data of the radioactive contamination resulting from the accident at the Chernobyl NPP, an estimation of the Russian population exposed to external doses was carried out (*Table 4.18*). This estimation is an important indicator of the existence of radiation hazard to human health. The analysis of the whole volume of collected material for the past decade makes it possible to evaluate the amounts of the main radionuclides discharged from Unit–4 of the Chernobyl NPP during the accident. The data listed in *Table 4.19* is preliminary since scientists of the international community do not yet fully agree on the magnitude of some radionuclide releases.

Table 4.19. Evaluation of the main radionuclide release resulting from the Chernobyl NPP accident.

Group of elements	Radio-nuclide	Half-life time	Activity, accumulated by the time of the accident, PBq	Evaluation of the discharge during the accident, PBq	
Noble gases	^{133}Xe	5.3 days	6,500	6,500[b]	6,500[c]
Volatile	^{131}I	8.04 days	3,200	1,760	1,200–1,700
	^{132}Te	3.26 days	2,700	1,150	1,000
	^{134}Cs	2.06 years	180	54	44–48
	^{137}Cs[a]	30.1 years	280	85	74–85
Intermediate	^{103}Ru	39.4 days	4,800	168	170
	^{106}Ru	367 days	2,100	73	30
	^{140}Ba	12.8 days	4,800	240	170
Refractory	^{89}Sr	50.6 days	2,300	115	81
	^{90}Sr	28.5 years	200	10	8
	^{95}Zr	64 days	5,600	196	170
	^{99}Mo	2.75 days	4,800	168	210
	^{141}Ce	32.5 days	5,600	196	200
	^{144}Ce	284 years	3,300	116	140
	^{238}Pu	86.4 years	1	0.035	0.03
	^{239}Pu	24,110 years	0.85	0.03	0.03
	^{240}Pu	6,553 years	1.2	0.042	0.044
	^{241}Pu	14.7 years	170	6	5.9
	^{242}Cm	162.8 days	26	0.9	0.93

[a] According to data presented in the *Atlas of Cesium Deposition on Europe after the Chernobyl Accident* (Israel, 1998), the quantity of ^{137}Cs deposited on the European territory was 70 PBq.
[b] Data from Devell *et al.*, 1995.
[c] Data from Borovoy *et al.*, 1992.

4.3 Handling Radioactive Material and Radioactive Waste at Nuclear Power Plants

The SNF from NPP reactors contains a significant amount of fissile isotopes; in breeders, the final amount of these isotopes is higher than the initial quantity, which is the physical basis of nuclear fuel breeding. From this point of view, spent fuel is another significant element of the raw material reserve of nuclear power and thus cannot be considered RW. This perspective is the official position of the Ministry of the Russian Federation for Atomic Energy (Minatom) and is shared by some other countries.

At present the only facility reprocessing the spent fuel from VVER–440, BN–350, and BN–600 is in operation in the town of Ozersk, in the Chelyabinsk region of Russia. It reprocesses all spent fuel both from these reactors of the FSU countries and from Soviet-designed NPPs in some East European states and Finland. The uranium separated in spent fuel reprocessing is used for producing new nuclear fuel for RBMK-type reactors. RBMK spent fuel is stored in special repositories at NPP sites. For economic reasons, no provision for reprocessing this type fuel has been made. At present, VVER–1000 spent fuel from Russian and Ukrainian NPPs is transferred for temporary storage to a special repository that is at the Krasnoyarsk Mining and Chemical Combine Site. The current capacity of the repository (6,000 tons) is expected to be completely filled by 2000.

Until recently, NPP RW has been handled by storing solid and evaporated liquid (still residue) wastes in special repositories at NPP sites. In the past decade, some RW conditioning technologies began to be introduced, because the method for SRW and LRW storage at the NPP could only be considered a temporary measure. Further development of nuclear power requires designing and implementing a comprehensive technical policy for RW handling. Such a policy should determine the development of systems for reprocessing, storage, shipment, and final disposal of RWs.

The final objective of the policy is to check and test the RW-reprocessing technologies and equipment, develop projects for their safe storage and disposal, and work out guides and standards on RW handling. Taking into account the actual conditions in Russia, the policy on NPP RW handling stipulates five stages at the NPP (*Figure 4.1*) and a final disposal stage.

At present, no Russian NPP has a complete set of facilities for RW conditioning. At some NPPs, liquid wastes are reprocessed in bituminous installations (Leningrad and Kalinin) or at deep evaporation facilities (NovoVoronezh NPP and Balakovo NPP). SRWs are compacted, upon sorting, at Beloyarsk, Kola, and NovoVoronezh NPPs, and combustible RWs are burned at Beloyarsk and Kola NPPs. At other NPPs, SRWs are stored without reprocessing. All facilities for waste conditioning are made in Russia; some of them need modernization.

Minatom has developed a RW handling program for the nuclear industry that includes a section on NPP RWs handling. This program and the NPP RW handling concept were used as the basis for development

Liquid Radioactive Waste | **Solid Radioactive Waste**

Collection, Classification, Sorting	• By type • By activity level • By category
Interim storage	• In tanks • In bunkers
Conditioning	• Concentration • Solidification • Burning • Grouting • Decontamination • Packing or containerization
Interim storage	• In surface structures • In near-surface structures
Transportation	• Formation and shipping of casks • Discharging of casks
Disposal	• In geological formation • In near-surface facilities

Figure 4.1. Scheme of NPP RW handling policy.

of the "Working Program on Radioactive Wastes Handling at NPPs of the State Concern Rosenergoatom," which addresses aspects of the NPP RW-handling problem, including the identification of tasks for the improvement of the NPP RW handling system, and the realization of these tasks.

These tasks include:

- Development of normative documents regulating RW handling operations.
- Construction of LRW grouting facilities.
- Construction of facilities for burning combustible RWs.
- Construction of SRW compaction facilities.
- Construction of facilities for vitrification of salt fusion cake and ashes.
- Development of technologies and equipment for removing salt cakes from LRW tanks.
- Development of technology and equipment for reprocessing filtering materials and slurry cakes.
- Development of technology for removing RWs from SRW storage compartments.
- Organization of process for sorting and transferring SRW for reprocessing.
- Construction of complex for metal waste reprocessing.
- Manufacturing containers for conditioned wastes.
- Designing and constructing storage facilities for reprocessed RWs.
- Designing and constructing dry repositories for SNF.

The realization of the main trends of NPP RWs handling requires completing R&D work on technologies and equipment for RW reprocessing and providing all NPPs with a complete set of these technologies and equipment.

Within the framework of the Working Program, research projects on the development of RW reprocessing technologies are being carried out; they aim at reducing waste in the final reprocessing stage (prior to storage) by concentrating activities as well as reusing some materials.

Unfortunately, because of financial difficulties in the nuclear industry and in Russia, only 15–20% of the Working Program has been completed during the first three years, mainly the testing of new technologies at the stage of RW conditioning. These developments have formed the basis for work on new RW-reprocessing facilities with Western companies. A weak point in the nuclear industry program and Working Program for nuclear power plants is the lack of advances in creating regional NPP RW burials, as this requires implementing expensive long-term storage of conditioned RWs at NPP sites.

4.4 Completeness and Reliability of Data on Radioactive Materials and Radioactive Wastes at Nuclear Power Plants

The data on the quantities of RM and RW at NPPs, given in Section 4.2 are complete. However, until now, the records on nuclear material in Russian nuclear power have been kept at each NPP in the form adopted at the NPP, with the exception of the RBMK NPPs, whose data on spent fuel are taken into account in the form of state statistics reports (form 3-TEKh).

Efforts are now underway to develop databases on nuclear materials at NPPs in accordance with a Presidential Decree of 15 September 1994, entitled "On Urgent Measures on Improvement of Accounting and Security of Nuclear Materials in the Utility Organization (Concern Rosenergoatom)."

The data on fission products activity in the spent fuel stored at the NPP sites are reliable, under assumptions (all fuel discharged from the reactors was considered to have the rated burn-up with a correction for the half-life made once a year) taken into account in the calculations. This characteristic can be improved after developing methods that take into account nuclear materials standard for all NPPs, and creating a central database for Concern Rosenergoatom.

Less reliable are the data on the activity of SRWs and LRWs stored at NPP sites. This is accounted for by the SRW classification method (by the gamma-radiation dose rate) adopted in the Soviet Union (and later Russia), normative documents on NPP safety, and by the methodical difficulties with determining the activity of LRWs with a high salt content. Given that the activity of the SRW and LRW stored at NPP sites is much lower than the fission product activity in spent fuel, the error in the determination of this type of RW are disregarded.

However, in the implementation of treatment and disposal methods, the absolute values are also important.

Bibliography

Abaguyan, A.A., Beresin, B.Ya., Vitkov, V.M., and Morozov, S.K., 1993, Experience of NPP operation in Russia and current goals, *Atomic Energy*, **74**(4):271.

Borovoy, A., 1992, Characteristics of the nuclear fuel of Power Unit No. 4 of Chernobyl NPP, In: *Radioecological Consequences of the Chernobyl Accident*, Kryshev, ed., pp. 9–20, Nuclear Society International, Moscow, Russia.

Buzulukov, Yu.P. and Dobrynin, Yu.L., 1993, Release of radionuclides during the Chernobyl accident, In: *The Chernobyl Papers*, S.E. Merwin and M.I. Balonov, eds., Research Enterprises, Richland, WA, USA, **1**:3–21.

Chernobyl design defects – the British view, 1987, *Nuclear Engineering International*, **32**(395): 21.

Chernobyl: Five Hard Years (collection of articles), 1991, Atompress, Moscow, Russia.

Commissariat a l'Énergie Atomique, 1993, *Les Centrales Nucléaires dans le Monde*, Commissariat à l'Énergie Atomique, Paris, France.

Consensus growing on positive scram at Chernobyl, 1988, *Nuclear Engineering International*, **33**(402):16-19.

Devell, L., Guntay, S., and Powers D.A., 1995, *The Chernobyl Reactor Accident Source Term: Development of a Consensus View*, Committee On the Safety of Nuclear Installations, (CSNI) report of the Nuclear Energy Agency (NEA)/Organization of Economic Cooperation & Development (OECD), Paris, France.

Four countries declare for nuclear power, 1991, *Nuclear Engineering International* **36**(442):10.

Gagarinski, A.Yu., 1989, Chernobyl today: State of research, *Nuclear Safety*, **30**(1):18–22.

Germany and USSR extend cooperation on safety, 1988, *Nuclear Engineering International*, **33**(413):3.

Gidrometeoizdat, 1990, *Chernobyl: Radioactive Contamination of Natural Environment*, Gidrometeoizdat, Moscow, Russia.

Grodzinski, M., 1991, Study of Chernobyl NPP accident consequences, *Radiobiologiya*, **31**(2)163–166.

Information on the accident at Chernobyl NPP and its consequences, 1986, *Atomic Energy*, **61**(5):320.

IAEA (International Atomic Energy Agency), 1989, *IAEA Incident Report NEA #6164*, September, Vienna, Austria.

IAEA (International Atomic Energy Agency), 1990, *IAEA Incident Report NEA #6188*, September, Vienna, Austria.

Israel, Yu.A., ed., 1991, *The Map of Radiation Situation at the Territories of the European Part of the USSR as of December 1990, 1:500,000*, by Israel, Yu.A., Kvasnikova, E.V., Nazarov, I.M., Fridman, Sh.D., Stukin, E.D., *et al.*, Northwestern Aerogeodetic Enterprise GUGK, Moscow, USSR.

Israel, Yu.A., ed., 1993a, *Map of Radioactive Contamination of the European Part and Ural Region of Russia by Cesium-137 as of January 1993, Scale: 1:500,000*, by Israel, Yu.A., Nazarov, I.M., Fridman, Sh.D., Kvasnikova, E.V., Stukin, E.D., Tsaturov, Yu.S., *et al.*, Institute of Global Climate and Ecology of Roshydromet and AS of Russia, Moscow, Russia.

Israel, Yu.A., ed., 1993b, *Map of Radiation Situation at the Territory of the European Part of CIS and the Baltic States as of January 1993; Contamination Level of Locality by Cesium-137; Scale: 1:2,500,000*, by Israel, Yu.A., Fridman, Sh.D., Nazarov, I.M., Kvasnikova, E.V., Stukin, E.D., Tsaturov, Yu.S., *et al.*, Institute of Global Climate and Ecology of Roshydromet and AS of Russia, Moscow, Russia.

Israel, Yu.A., Kvasnikova, E.V., Nazarov, I.M., and Fridman, Sh.D., 1994, Global and regional radioactive contamination of the European territory of the ex-USSR by cesium-137, *Meteorology and Hydrology*, **5**:5–9.

Israel, Yu.A., Stukin, E.D., and Tsaturov, Yu.S., 1996, Dynamics of Chernobyl radionuclides behavior in the environment (report from conference), *One Decade After Chernobyl: Summing up the Consequences of the Accident*, 8–12 April, Report CN-63/386, Austria International Center, Vienna, Austria.

Israel, Yu.A., ed., 1998, *Atlas of Contamination of the European Part of Russia, Belorussia and Ukraine*, Institute of Global Climate and Ecology, Roshydromet, Roskartographiya, Moscow, Russia.

Israel, Yu.A., 1998, *Atlas of cesium deposition on Europe after the Chernobyl accident*, Contributing organizations: Institute of Global climate and Ecology, Moscow, Russia; Environment Institute, EC Joint Research Centre, Ispra, Italy; Roshydromet, Moscow, Russia; Belhydromet, Ministry of Emergencies, Minsk, Belorussia; and Vinchernobyl, Derzhkomhydromet, Kiev, Ukraine; European Commission, XII.F.6, Brussels, Belgium.

Konstantinov, L.V. and Gonzales, A.J., 1989, The radiological consequences of Chernobyl accident, *Nuclear Safety*, **30**(1):53–69.

Lithuania and Sweden identify upgrades for Ignalina, 1992, *Nuclear Engineering International*, **37**(453):2.

Logachev, V.A., 1996, Assessing and comparing the population dose rates of the Russian Federation after nuclear tests in the atmosphere at the testing sites of the ex-USSR and after the Chernobyl NPP accident (report from conference), *One Decade After Chernobyl: Summing Up the Consequences of the Accident*, 8-12 April, Austria International Center, Vienna Austria.

Minergo (Ministry of Energy), 1985, Model content of the plan of measures for personnel protection in the case of radiation accidents at NPPs, Minergo, Moscow, USSR.

NRER (National Radiation and Epidemiological Registry), 1995, Radiation doses for emergency workers, In: *Radiation & Risk, Special issue of the Bulletin of NRER*, **95**(2), Obninsk-Moscow, Russia.

NRSRB (Nuclear and Radiation Safety Regulatory Body), 1995, *Report on the 1995 Activity of the NRSRB of RF*, RD-03-02-95, NRSRB of RF, Moscow, Russia.

Obninsk Institute of Atomic Power, 1992, *Accidents and Incidents at Nuclear Power Plants* (manual), Obninsk Institute of Atomic Power, Obninsk, Russia.

Positive scram: the key to Chernobyl or only a contributor?, 1987, *Nuclear Engineering International*, **32**(399):9-10.

Radiation situation at the territory of the Russian Federation (from the state report about the environmental situation of the Russian Federation in 1993), 1995, *Bulletin of TsNIIAtominform*, pp. 3–4.

Reznik, A.A., Dmitriev, S.A., and Medvedev, S.A., 1996, Paper presented to the coordination research council on RW handling, *Atomic Energy*, **81**(1).

Rippon, S., 1991, Five years, uncertainties remain at Chernobyl, *Nuclear News* **34**(8):49–52.

Sidorenko, V.A., 1994, Conceptual aspects of nuclear power development in Russia up to 2000, *Atomic Energy*, **76**(4):259.

The Chernobyl accident, 1987, *Nuclear Safety*, **28**(1):6–9.

Thomsom, B., 1991, IAEA signs agreement on Chernobyl Centre, *Nuclear Engineering International*, **35**(436):10.

Thomsom, B., 1991, International effort needed on sarcophagus, *Nuclear Engineering International*, **36**(446):17-19.

TsNIIAtominform, 1992, *Information Bulletin, Special Issue* (21 December), TsNIIAtominform, Moscow, Russia.

Varley, J., 1990, Lessons from insider's notebook, *Nuclear Engineering International*, **35**(431):50-51.

Whipple, C. and Starr, C., 1988, Nuclear power safety: Goals in light of the Chernobyl accident, *Nuclear Safety*, **29**(1):20–26.

Was pump cavitation the key to Chernobyl?, 1988, *Nuclear Engineering International*, **33**(408):18-19.

Yegorov, N.N., Kudryavtsev, E.G., Nikipelov, B.V., et al., 1993, Regeneration and localization of radioactive wastes and of NPP nuclear cycle, *Atomic Energy*, **74**(4):307.

Chapter 5

Naval Application of Nuclear Power

5.1 Introduction

5.1.1 Objects of the Russian Navy and the Ministry of Economy

The main sources of radiation hazards and environmental contamination from the Russian Navy and from objects of the Department of Defense Industry of the Ministry of Economy include:

- Submarines and surface ships with ship propulsion reactors (SPRs).
- Technical vessels supporting nuclear-powered ships: technical tankers, technical depot ships for reloading of reactors, and *Amur*-type technical tankers.
- Naval bases of the ships with SPRs.
- Shipyards where outdated ships with SPRs are stored.
- Temporary storage facilities for fresh and spent nuclear fuel (SNF), including coastal bases and depot ships, and depot technical ships, which reload nuclear ship reactors.
- Ship-repairing and ship-building factories of the Ministry of Economy, where repairing of nuclear ships and reloading of the reactor core take place.
- Radioisotope energy sources.

Radiation hazard may be the result of tests of nuclear and radiological weapons and malfunctions associated with them. In addition, reactor operations and dismantling at the Paldisky (Estonia) training center and reactor operations at the Sevastopol Higher Naval Engineering School (at present belonging to the Ukraine) are areas of concern.

As of early 1995, about 140 nuclear submarines (NSs) had been taken out of operation. By 2000, 156 NSs will have been removed from the Navy stock, including 95 NSs and 61 NSs from the Northern and the Pacific Fleets, respectively.

In the Russian Federation (RF), a program for dismantling NSs has been adopted and is being implemented. However, the capacity of the program to accomplish these works is far from sufficient. The mass withdrawal of second-generation NSs from operation, in accordance with the international large-scale commitments of the RF under disarmament agreements, has coincided with the planned withdrawal of first-generation NSs from operation because of wear and tear. As a result, there is a large queue of NSs waiting to be decommissioned. The first delay is waiting for SNF removal; then, cutting out, sealing, and storing of the reactor compartment is undertaken until ionizing radiation levels are reduced due to radionuclide decay to a level that permits subsequent work on reactor dismantling. At this point in the process, the long-term storage of reactor compartments has not yet begun.

The great number of decommissioned NSs with fuel-less reactors and reactors containing SNF, along with the increasing number of isolated reactor compartments, has created a problem where radiation safety must be ensured under conditions quite different from those of normal NS operation.

At present, up to 66% of decommissioned NSs are stored with SNF onboard. The stored ships are dispersed geographically, and the NSs are being gradually sent to ship-repairing yards. The enhanced dispersion of NSs is evident in Severodvinsk. In March 1995, 16 NSs, to be utilized and repaired with 32 nuclear reactors still containing nuclear fuel, were within the precincts of this town with a population of 240,000.

Since 4 July 1992, a pilot program for decommissioning and conversion of 13 NSs has been under way with the following distribution of submarines by four ship-repairing yards:

Industrial Association (IA) Sevmashpredpriyatie (Severodvinsk)	3
State Machine-building Enterprise (GMP) Zvezdochka (Severodvinsk)	4
Far-East Plant (DVZ) Zvezda (B. Kamen, Primorsk land)	4
Ship Repair Plant (SRZ) Nerpa (Snezhnogorsk, Murmansk region)	2

The work was completed in 1995–1996.

There is a strong tendency for the accumulation of SNF in Severodvinsk, because the actual Navy capacity for SNF discharge in the Northern region is not more than three NSs per year. The liquid radioactive wastes (LRWs) reprocessing has not been balanced yet: LRW accumulation is about 2,000–2,500 m^3/yr, whereas the capacity of the reprocessing installations is somewhat more than 1,000 m^3/yr.

In the Far East, NS decommissioning/conversion has been affected at DVZ Zvezda, where the capacity has been two to three NSs per year instead of the assumed capacity of five to six NSs per year.

Earlier, 20 cores were delivered annually to the reprocessing plants. Currently, because of the changeover to new transport containers such as the TUK–18, the shipping capabilities have been reduced in part due to the insufficient quantity of new containers and other reasons.

5.1.2 The civilian marine fleet

At present, eight civilian nuclear ships are in operation in Russia: seven nuclear icebreakers and one lighter, *Sevmorput*, belonging to the Murmansk Marine Steamship (MMS) line. The first nuclear icebreaker, *Lenin*, was put out of service in 1990. Its SNF was unloaded from the reactors. Other data about activity and radionuclide content for SPR equipment, as well as circuit surfaces of ship reactors, are unavailable. *Lenin* is in an Atomflot base, which is engaged in repairing the reactor facilities of MMS; storing the SNF; and collecting, storing, and reprocessing the radioactive wastes (RWs) generated during the ship's operation.

The decommissioned floating technological base, *Lepse*, which serves as a storage facility of defective SNF, has been laid up here since 1988.

Table 5.1. Characteristics of RWs stored by the Russian Navy and the Ministry of Economy as of 1 January 1994.

Enterprise	Quantity of LRW, 10^3 m^3	Total activity of LRW, Bq	Quantity of SRW, 10^3 m^3	Total activity of SRW, Bq
Russian Navy				
Northern Fleet	8.695	3.05×10^{12}	5.863	5.41×10^{13}
Pacific Fleet	5.767	1.36×10^{12}	1.703	6.29×10^{14}
Ministry of Economy				
JC Amur Shipbuilding Enterprise	1.0	0.5×10^{9}	0.012	n.a.
SRZ Nerpa	0.105	1.11×10^{10}	0.5	n.a.
DVZ Zvezda	0.95	1.78×10^{11}	1.99	4.7×10^{12}
IA Sever	1.033	n.a.	5.76	5.44×10^{12}
IA Sevmashpred-priyatie	0.123	n.a.	1.8	n.a.
Ship Equipment Enterprise	0.02	3.7×10^{5}	0.0015	0.67×10^{10}
Total	17.693	4.6^{12}	17.63	6.91×10^{14}

n.a. = not available.

5.2 Characteristics of Accumulated Radioactive Materials and Radioactive Wastes

5.2.1 Quantity, composition, storing, and reprocessing conditions of radioactive material and RW

The data on RWs stored by the Russian Navy and factories of the Department of Defense Industry of the Ministry of Economy are presented in *Table 5.1*. More recent data have not been published.

Solid radioactive wastes (SRWs) are present in construction materials of the nuclear reactors, fuel elements of the damaged submarines, spent ion exchange resins used in water purification filters of the primary circuit, and radioactive sediments formed during SPR accidents.

In NS reactors without SNF, the radionuclides of two groups are potentially hazardous: radionuclides of the activation origin and radioactive fission products. In the first radionuclide group, the hazard is determined by accidental spread of ^{55}Fe (half-life is 2.7 years), ^{60}Co (5.27 years), and ^{63}Ni (96 years), and sometimes ^{152}Eu (13.3 years), ^{154}Eu (8.8 years), and ^{155}Eu (4.96 years). In the second group, the main responsible radionuclides are ^{90}Sr (29 years), ^{134}Cs (2.06 years), and ^{137}Cs (30 years). Europium radionuclides are formed in the control rods of some nuclear reactors, although in some designs there is no europium in the control rods. Fission products may be present in the reactor coolant circuit in the form of sediments on the inner surfaces as a result of the reactor operating with failed (nonsealing) fuel rods.

Decommissioned NSs differ by modes of previous operation. The most conservative supposition (operation at power equal to 20% of nominal for 200,000 hours) is appropriate to the accumulation of the following radionuclides in metal structures inside the reactor within two years after the reactor is shut down: ^{55}Fe about 1.5×10^5 Bq, ^{60}Co about 1.4×10^5 Bq, and ^{63}Ni no more than 0.3×10^5 Bq. In the conservative supposition, one would expect to find up to 1.9×10^{11} Bq of long-lived ^{90}Sr and ^{137}Cs in the sediments inside the primary circuit.

The unloading of NSF in NSs is accompanied by the generation of LRWs and SRWs, of which LRWs are the most mobile and the most potentially hazardous. The total amount of LRWs accompanying the unloading of one reactor is about 50 m^3 at a radionuclide concentration not exceeding, as a rule, 3.7×10^5 Bq/l. Such LRWs belong to the low-level RW category.

LRWs produced by Navy installations have specific activities from 3.7×10^3 to 3.7×10^8 Bq/l, of which about 30% is due to slightly salty drainage waters of the circuits and 70% is due to highly salty decontamination waters from operating and repairing SPRs. Depending on the design of the submarine, the reloading of one NS reactor produces from 100 m^3 to 150 m^3 of LRWs on average, of which approximately 50% have specific activity that is more than 3.7×10^5 Bq/l, including about 10% that have activity higher than 3.7×10^6 Bq/l. The volume of decontamination water is about 60 m^3; 8–10 m^3 of the water has a specific activity of more than 3.7×10^5 Bq/l.

LRWs produced in the process of operating, repairing, and utilizing the Navy nuclear ships are stored in coastal and floating tanks, on floating technical bases for reloading the NS reactors, and on the technical tankers. The existing SNF and RW storage facilities were built in the 1960s.

The Navy possesses the following components for storing and reprocessing the LRWs:

- The Northern Fleet has coastal storages of 5,300 m^3 total capacity, floating storages of 3,700 m^3 total capacity, and a station for reprocessing the radioactively contaminated water of the technical transport *Amur*.
- The Pacific Fleet has coastal storages of 3,500 m^3 total capacity, floating storages of 3,000 m^3 total capacity, the radioactively contaminated water-reprocessing station of the technical transport *Pinega*, and two experimental mobile plants, *Sharya-M* (in experimental operation since August 1994).

At present, up to 30% of coastal and floating tanks are not in service because of expired operational guarantees or the lack of watertightness. The remaining 70% of coastal and floating reservoirs are nearly completely filled. Their construction is inconsistent with international recommendations, as well as national and departmental requirements for ensuring radiation safety for storing LRWs. Temporary storage facilities for RWs have not been built at the fleets because financing is lacking.

The Northern and Pacific Fleets have one ship each (the *Amur* and *Pinega*), for RW transportation and reprocessing. Each ship carries about 800 tons of LRWs and has a reprocessing station for radioactively contaminated waters that is designed to purify up to 5 m^3 of water per hour with a decontamination factor equal to 100. These facilities are not operating, as there is no equipment for solidification of radioactive sludge, a product of LRW reprocessing. Specifications for modernizing the ships' LRW-reprocessing stations have been developed. Moreover, three ships (two in the Northern Fleet and one in the Pacific Fleet) that can reload the NS reactors were also designed to purify the primary coolant with a capacity of 10 m^3/hr. However, these systems are not used because of technological drawbacks.

The LRW reprocessing in the Northern Fleet is temporarily carried out at the Russian Transport Enterprise (RTE) Atomflot (in 1994 and 1995 it received 1,000 m^3 of LRWs to reprocess). To increase the capacity of the Atomflot in Murmansk, a Russian–Norwegian–American project was started in 1996. The upgrade of the treatment facility will increase capacity from 1,200–5,000 m^3/yr. This will cover the needs of the icebreaker fleet, will permit the processing of high-salt waste from the Russian Navy's Northern Fleet, and will improve stabilization and interim storage of the processed waste.

The LRW reprocessing in the Pacific Fleet is performed by two experimental facilities of *Sharya-M* type, which have a capacity of 1.0–1.5 m^3 per day each. The annual volume of LRW generated from decommissioning and dismantling of nuclear submarines in the Russian Far East is estimated to be approximately 7,000 m^3. Currently, there is no treatment facility in the Russian Far East that could handle this waste. To avoid ocean dumping of LRW, a joint Russian–Japanese project was launched for constructing a processing facility for RW from nuclear submarines in the Russian Far East. The treatment facility, designed to be mounted on a barge, will be moored at the quay of a shipyard near Vladivostok. The cost of the project is US$ 25 million.

All SNF from the bases of the Northern and Pacific Fleets is scheduled to be removed for reprocessing at facilities of the Ministry of the Russian Federation for Atomic Energy (Minatom). Not all SNF (such as damaged SNF) will be reprocessed immediately; some will be stored for later disposal.

SNF used in the reactors of the training center situated in the town of Paldisky has been removed completely, and the conservation of other reactors' RW at the site, including that from SPRs, has also been completed.

At present, the problems of handling the SNF not subject to reprocessing, as well as the fuel from damaged NS reactors, have not been solved.

5.2.2 Accidents with Navy SPRs

The accidental ingress of seawater into the primary circuit of a SPR is accompanied by the destruction of radioactive deposits on the inner surfaces of the reactor (which accumulate during operation) and the subsequent escape of radionuclides into the water. At the release of radioactive materials (RMs) into the environment, the content of these sediments makes a contribution comparable with that of radionuclides from metal structures, although the absolute quantities of nuclides accumulated are quite different. Higher by some orders, the ^{60}Co content in the vessel structures proves to be comparable in radiation hazard to the release of small quantities of ^{90}Sr and ^{137}Cs from the sediments. In this case, the main factor is the rather low rate of metal corrosion in seawater, which is advantageous because the ^{60}Co content can then decay to less hazardous levels. The accidental release of radionuclides to seawater due to corrosion is estimated at 6.29×10^{10} Bq/yr for ^{55}Fe and ^{60}Co releases and 1.48×10^{10} Bq/yr for

^{63}Ni releases. Altogether, 4.81×10^{11} Bq of ^{60}Co, 4.44×10^{11} Bq of ^{63}Ni, and approximately 1.85×10^{11} Bq of ^{90}Sr and ^{137}Cs would enter the water during the post-accident period. Radiation doses produced in this case are not hazardous from the standpoint of radiation impact on environmental objects.

A more serious situation is caused by the accidental sinking of NSs with reactors containing SNF. In these situations, there is the potential hazard of the release into seawater of the radionuclides accumulated in the fuel of the reactor. For a period of two years after the shutdown of a NS reactor, the total radionuclide radioactivity in SNF is characterized by a level of about 7.4×10^{15} Bq, with the following radionuclides making the main contributions: ^{144}Ce (2.2×10^{15} Bq); ^{147}Pm (1.2×10^{15} Bq); ^{134}Cs, ^{137}Cs, and ^{90}Sr (0.4×10^{15} Bq each); as well as ^{106}Ru (0.2×10^{15} Bq). Both ^{144}Ce and ^{106}Ru have relatively short half-lives: 0.78 years and 1 year, respectively. The half-lives of ^{134}Cs and ^{147}Pm are somewhat longer: 2.06 years and 2.64 years, respectively. ^{90}Sr and ^{137}Cs have the longest half-lives – about 30 years. Their radioactivity in the SNF is 2,000 times higher than was predicted for possible surface sediments in the SPR with unloaded fuel. This indicates a much greater radiation hazard for sunken NSs with SNF in the reactor than those without spent fuel.

It follows from the experimental data obtained in Murmansk at the Atomflot base that the rate of the ^{90}Sr release from the fuel rods to the seawater will not exceed 1% each year from the activity accumulated in the SNF. For designed burn-up, the ^{90}Sr release (3.7×10^{12} Bq/yr) is 20 times higher than the previous estimate for the reactor sinking without SNF.

On the whole, it is reasonable to assume that accidents associated with NSs sinking without SNF aboard do not cause any serious radiation consequences. In parallel with this, additional quantitative analyses is needed of the radioecological consequences of the sinking of NSs with SNF in the reactors. The conservative approach commonly used for such calculations do not take into account fully the complexity of migration paths of radionuclides from their source (the fuel rods with SNF) to the environment (the sea around the sunken NS). There is no reason to assume that the seawater ingress to the reactor gives rise to rapid mass cladding failure under conditions where there is large-scale interaction of the seawater with the fuel rod kernels. Through experiments, it has been determined that corrosive destruction of the fuel rods does not create an extensive zone of direct contact of the pellets with water. In fact, radionuclides penetrating into the primary circuit of the reactor are brought into stagnant water. Its exchange with the environment cannot be active; it takes place with the retardation of the nuclides first in the reactor volume, then in the reactor unit equipment compartment, and finally in the volume outside this compartment. Precipitation of the nuclides occurs in all these volumes due to their transport retardation. Five-year, full-scale observations of ^{137}Cs releases from the reactor of the sunken NS *Komsomolets* using highly sensitive equipment indicate a low rate of radionuclide transport out of the NS hull (no more than 3.7×10^{10} to 3.7×10^{11} Bq/yr).

Along with the accidents associated with NS sinking and the penetration of seawater into the SPR, reactivity accidents resulting in initiating the fission chain reaction should also be mentioned. However, when the NS is laid up, any potential reactivity changes in the reactors with fuel are blocked by fixing the positions of all operation units by reliable mechanical means. SPRs of NSs withdrawn from operation are under deep subcritical conditions.

The Russian Research Center–Kurchatov Institute (RRC-KI) and Navy archives have relatively detailed information available for only one NS accident that released considerable radionuclides to the environment. This accident took place on 10 August 1985 on one of the Pacific Fleet's submarines in the Chazhma Bay (Shkotovo–22, Primorsky land). After completing the reloading of the reactors, a spontaneous chain reaction was initiated in the left board reactor because of a violation of nuclear safety requirements regarding the removal of the reactor's cover. In response to the thermal explosion, the bow and the after-body apparatus compartments were destroyed. Part of the newly loaded fuel was thrown from the reactor's vessel. The reloading facility was demolished (its roof was thrown 70–80 m by the explosion and landed in the water 30 m from the coast). The submarine suffered damage to the hull in the after-body part of the reactor compartment. Immediately after the explosion, a fire ignited in the reactor compartment; the fire was contained in four hours.

Investigators found that combustion products – together with both fission and activation products and the particles of unreacted fuel composition (in the form of coarse-dispersed particles and slag) – fell around the damaged submarine in a radius of 50–100 m. Seven and one-half hours after the accident, the exposure dose rate (EDR) in the area reached

250–500 mrem/hr, with a surface density of radioactive contamination of $(0.1–3.6)\times10^5$ Bq/cm^2.

The gaseous radioactive cloud from the accident crossed the Peninsula Danube in a northwest direction and traveled to the Ussuriysky Gulf. Full-scale studies of seawater and bottom sediments showed that, as the radioactive cloud moved over Ussuriysky Gulf (width of 28–30 km), the density of fallout decreased to background levels and did not influence the radiation situation in Vladivostok and the coastal zone. The EDR near the coastline of the Shamora Bay was equal to 6–8 mrem/hr from 1989 to 1991.

Weather conditions at the time of the accident were favorable: The Shkotovo–22 settlement located 1.5 km from the submarine's mooring did not suffer from radioactive fallout. The bulk of the radioactivity fell in the territory adjacent to Chazhma Bay where a factory is located and in the uninhabited part of the Peninsula Danube.

Estimates show that the release of the short-lived radioactive noble gases totaled approximately 7.4×10^{16} Bq. The other volatile fission (basically, the iodine isotopes) and activation (predominantly, ^{60}Co and ^{54}Mn) products totaled about 1.85×10^{17} Bq.

The radiation monitoring of the environment revealed that the highest density of fallout was concentrated in a forest tract on a narrow area 3.5 km long and 200–650 m wide with a total area of 2 km^2. In five to seven months, the radiation situation was normalized over the whole territory of the factory. Two months after the accident, the radionuclides content in seawater had decreased to initial background levels.

The radioactive trace on the Peninsula Danube was due primarily to ^{60}Co activity (95–99%), and, to a considerably smaller degree, was due to ^{137}Cs activity. This trace is highly nonuniform and concentrated in patches. In one of the trace sections, a temporary near-surface storage facility for the contaminated factory equipment was built. The storage and the main part of the trace on the uninhabited territory were fenced with barbed wire. A long-term analysis of the measurement dynamics of environmental radioactive contamination shows that radionuclides are primarily strongly sorbed into the soil. Their total removal with the surface waters to the Chazhma Bay is assessed as $1.85\times10^8–7.4\times10^{10}$ Bq/yr.

As a result of the accident, the long-lived radioactive contamination of the Chazhma Bay bottom sediments was formed primarily by ^{60}Co (96–99%) and partially by ^{137}Cs. The area of radioactive contamination is about 0.1 km^2. At the center, the EDR reaches 20–40 mR/hr (maximum 117 mR/hr as of 1992).

The maximum specific activity of ^{60}Co in the bottom sediments at the place of the accident amounts to 7.77×10^4 Bq/kg, and in the sea biota it amounts to as much as 6.66×10^5 Bq/kg. Movement of the contaminated bottom sediments from the Chazhma Bay to the western passage of the Arrows Gulf has been observed. Contamination of the eastern part of the Ussuriysky Gulf, in a radius of 3–5 km from the place of the coastal radioactive trace, causes the EDR to be above background levels within the limits of 1–8 μR/hr. The concentrations of radionuclides in the Chazhma Bay, the western passage of the Arrows Gulf, and the eastern part of the Ussuriysky Gulf are at the level of background values, characteristic of other regions along the Pacific Coast.

The observable tendency for the transport of radioactive contamination in the natural layer and its dispersion over the Chazhma Bay will not lead to serious ecological consequences, as the general activity of ^{60}Co in the bottom sediments is comparatively low (about 1.85×10^{11} Bq).

The radiation safety service of the factory carries out constant monitoring of the radioecological situation in the accident area and on the radioactive trace. The chemical and medical services of the Pacific Navy, seaside fleets, regional sanitary and epidemiological stations, and hydrometeorological organizations carry out periodic measurements of radionuclide contents in environmental objects.

The results of the full-scale observations and numerous radioecological inspections indicate that the accident on the submarine in the Chazhma Bay in 1985 has not had a measurable radiation effect on Vladivostok, its beach zone, or the Shkotovo–22 settlement. The residual long-lived radioactive pollution of the environment and bottom sediments in the Chazhma Bay region has been reliably localized and will not lead to serious radioecological consequences.

At present, the Navy has four damaged submarines, three of which are in the Far East in the Pavlovski Bay (project 675, serial No. 175 and 541; and project 671, serial No. 610) and one is in the North (project 675, serial No. 533). The cores of submarines No. 541 and No. 533 are to be discharged.

Technical measures for ensuring nuclear and radiation safety are provided during NS storage. However, the risk of accidental release of RMs into the environment is not negligible. In this connection, the information

Table 5.2. Characteristics of RWs stored at the RTE Atomflot.

Quantity of LRW, m^3	Total activity of LRW, Bq	Quantity of SRW, m^3 (t)	Total activity of SRW, Bq
436	5.61×10^{13}	730 (180)	3.77×10^{16}

about the reactor compartments as potential sources of radionuclides is very important.

Taking into account the work performed at the first stages of NS decommissioning, it is reasonable to assume that NSs with discharged SNF should be stored not less than two years after the shutdown of the nuclear reactors. Once the NS is stored, only long-lived nuclides are a danger to personnel, because short-lived nuclides have already decayed during storage (see Section 5.2.1).

5.2.3 Quantity, composition, storage, and reprocessing conditions of RW of the civilian marine fleet

The radioactivity of wastes produced in the operation of civilian icebreakers with SPRs and then temporarily stored on special crafts and in coastal RTE Atomflot facilities consists of both neutron activation and fission products.

Integrated data about RWs stored at the civilian nuclear fleet objects as of late 1995 are listed in *Table 5.2.* More recent data have not been published in the open press. The information is presented in annual reports of the Radiation Safety Service of MMS, which are regularly delivered to the Department of Marine Fleet of the RF Transport Ministry.

As pointed out, most of the RW is on the floating base *Lepse*, which contains damaged SNF. As of 1 January 1995, total activity of the stored material amounted to about 2.78×10^{16} Bq, with the percentage of ^{137}Cs and ^{90}Sr accounting for 70% and that of actinides (^{238}Pu, ^{239}Pu, ^{241}Am, and ^{244}Cm) being no more than 3% of the total activity. Solving the problems of the *Lepse* base was incorporated into the "Federal Program on RW and Spent Nuclear Materials Handling, Their Utilization, and Disposal for a Period from 1993 to 1995 and up to the 2005." Moreover, the problem outlined is reflected in the plans of activities of the Euroarctic (Barents) region.

Because of the frequency of ship movements in the Kola Bay, the possibility of ships colliding with the *Lepse* base cannot be ruled out. In addition, there is a possibility of the *Lepse* sinking because of its aging hull. This may produce a severe hazardous contamination of the environment. The development of solutions for the complex handling of the *Lepse* base is also of interest from the viewpoint of handling Navy auxiliary ships and compartments with SPR with unremovable SNF.

A sorption installation is operating in Murmansk on the RTE Atomflot. Designed for the decontamination of LRW from nuclear icebreakers, the installation has a capacity of 2.5 m^3/hr. This is the only installation in the country capable of decontaminating LRW of all types formed in operation of the SPRs to levels of radionuclide content within the requirements of Norms of Radiation Safety (NRS–76/87).

To the present, more than 8,000 m^3 of LRW of different types have been decontaminated at the special water treatment installation of the RTE Atomflot. The capacity of this installation permits an essential part of LRW of low-specific activity formed in operating and repairing the Northern Navy's ships and the MMS nuclear ships to be reprocessed. A complete solution to the problem on decontamination of all LRW produced in the Northern Fleet requires additional tools at the existing water treatment installation.

5.2.4 Marine RW disposal and sunken nuclear ships

SRW of low-and medium-specific activity dumped into the sea are enclosed in containers or, if not sealed, are composed of polyethylene films, clothes, footwear, tools, and faulty equipment accessories such as valves, pipes, circulation pumps, steam generators, and ion exchange filter casings contaminated in the repair process. Most radionuclides are concentrated in the equipment and its corrosion films. These are primarily activation products, such as ^{60}Co. The films also contain some fission products. This mixed composition of radionuclides is typical for SRW. It is impossible to separate them and give an adequate quantitative evaluation of each group. It is inexpedient to further determine activity and radionuclide content in connection with the small contribution of SRW to the total quantity of sunken wastes, as well as with the local character of water masses and bottom sediments. In the immediate vicinity of sunken SRW, this was specifically demonstrated by the results of international Russian–Norwegian expeditions to the Kara Sea in 1992–1995.

Table 5.3. Activity of long-lived radionuclides in the SPRs that have sunk in the vicinity of Novaya Zemlya as of late 1994.

Submarine or ship inventory number	Fission products, TBq	Activation products, TBq	Actinides, TBq	Total, TBq
Sunken reactors with fuel				
285	634	12.80	8.13	654.93
901	718	5.96	3.44	727.40
421	287	2.88	2.84	292.72
601	375	239.00	1.25	615.25
Total	2,014	260.64	15.66	2,290.30
Sunken reactors without fuel				
254		9.47		9.47
260		5.07		5.07
538		4.51		4.51
Total		19.05		19.05
Nuclear icebreaker Lenin				
OK-150	1,880	230	85.10	2,195.10
Total	3,894	509.69	100.76	4,504.45

The analysis and generalization of the information characterizing RW in the Arctic regions has shown that the long-lived radionuclides contained in SNF and SPRs at the bottom of the bays of the eastern coast and in the vicinity of the Novaya Zemlya archipelago may be the maximum potential radiation hazard.

For reactors, both with and without SNF, most of the information was obtained by the Radleg project. In addition to the materials listed in *White Book–93* in the framework of the International Arctic Sea Assessment Project (IASAP) International Program, the experts from RRC-KI and the Institute of Energy Physics (Obninsk) of Minatom have estimated the radionuclide composition and activity in all 10 SPRs that have sunk, as well as in seven containers of SNF that have sunk. It has been shown that the main part of activity is in the container with SNF of the nuclear icebreaker *Lenin*: It is comparable to the activity of all other sunken reactors including the reactors of damaged NSs with SNF (*Table 5.3*).

It has been also established that essential mistakes were made in the first official document, e.g., gross activity of sunken RW was overestimated by 2.5 times. Taking into account the decay of relatively short half-life radionuclides, this activity has never exceeded 2.59×10^{16} Bq. At present, the gross activity of radionuclides in the sunken SPR and SNF does not exceed 4.81×10^{15} Bq. It is many times less than the gross activity of long-lived radionuclides that were sunk by the states of Western Europe in the North Atlantic during the implementation of the London Convention of 1972.

After publishing the data about sunken NSs with SNF and other RW in the northern seas in 1993, the open inspection of gulfs of the Novaya Zemlya archipelago began. The first expedition was organized by the RF Navy on the *V. Kavraysky* in August and September 1993. Specialists from the Khlopin Radium Institute, the RRC-KI, the Navy, and the central testing site of the Ministry of Defense took part in the expedition. The gulfs of Ambrosimov, Stepovoi, Tsivolki, Oga, Sedov, Techenii, and Blagopoluchie, as well as Novaya Zemlya's depression, were inspected. In each gulf, two to three stations were set up to measure dose rates on the water surface, at different depths and the near bottom, as well as to determine the radionuclide activity by immersed gamma-spectrometer and by sampling the water and sediment deposits for radiochemical analyses. The investigations performed did not reveal any deviation from the background radiation due to natural radionuclides and global fallout.

Between 1993 and 1995, the international Russian–Norwegian marine expeditions inspected the above-mentioned Novaya Zemlya's gulfs, as well as Novaya Zemlya's depression. A correlation between the data obtained from this expedition and the results of a Navy expedition shows reasonable agreement. The general conclusion is that the radiation doses are low here, and that the radionuclide concentrations outside the gulfs are no different from concentrations in open regions of the Kara Sea. Higher concentrations of ^{137}Cs and ^{152}Eu revealed in some sections of the Stepovoi and Ambrosimov gulfs are not considered a function of the radionuclide releases from both sunken NSs and containers with SNF.

5.2.5 Radiation situation in the area of the wreckage of the Komsomolets nuclear submarine

In 1989 in the Sea of Norway, the *Komsomolets* NS sank to a depth of about 1,700 m with a propulsion nuclear reactor and two rockets with nuclear warheads aboard. Before the accident, the SPR of *Komsomolets* produced about 580 GWhr. For this rate of nuclear fuel burnup there is the following gross activity of both long-lived fission and activation products

Table 5.4. Long-lived artificial radionuclides in the North Atlantic, 10^{16} Bq.

Radionuclides	Global fallouts	Radiochemical plants	Komsomolets NS
^{137}Cs	7.6	3.0	0.31
^{90}Sr	5.1	4.3	0.28
$^{239+240}Pu$	0.13	0.06	0.0021

accumulated in the reactor core: ^{137}Cs 3.1×10^{15} Bq, ^{90}Sr 2.8×10^{15} Bq, ^{239}Pu 4.4×10^{12} Bq, and ^{240}Pu 1.7×10^{12} Bq.

Two nuclear warheads with activity of 1.48×10^{13} Bq are also aboard the NS. Thus, the gross activity aboard of $^{239+240}Pu$ is 2.1×10^{13} Bq.

The comparison with other sources of the anthropogenic long-lived radionuclides in the North Atlantic (*Table 5.4*) shows a relatively low input of radioactivity from the *Komsomolets* for the sea: The activity of ^{137}Cs, ^{90}Sr, and $^{239+240}Pu$ from the NS is ten times lower than that from global fallout and discharges of West European radiochemical plants as well as RW that sank earlier.

From 1991 to 1995, the Special Design Bureau of Marine Engineering (SDBME) Rubin together with other organizations (Khlopin Radium Institute, RRC-KI, etc.) organized a number of marine expeditions aimed at obtaining information about the radiation situation in the region of the wreckage in connection with possible radionuclides released into the sea.

The critical analysis used methods of concentrating radionuclides from large volumes of seawater (including using selective sorbents) and procedures of gamma-spectrometric and radiochemical analysis of the sea water, along with samples of bottom sediments and biota. The reliability of the results obtained has been confirmed by taking into account the extreme heterogeneity of the fields of radionuclides of artificial and natural origin in the inspected region.

According to calculations, it was found that the NS and SPR had lost sealing due to a collision with the seafloor, and, consequently, radioactive substances gained access to the environment. The rate of the radionuclides released depends on the character of the destruction of the primary circuit reactor, the ship structures, residual heat, the state of the nuclear warheads, and other conditions.

Fission products (^{144}Ce, ^{137}Cs, ^{90}Sr, ^{106}Ru) and nuclear activation products of ^{60}Co, ^{59}Co, ^{54}Mn, and ^{3}H, as well as plutonium, are the main radionuclides determining the radiation-ecological situation in the area of the NS wreckage.

The results of field work, executed between 1991 and 1995, showed the release of insignificant quantities of a number of radionuclides of artificial origin from the damaged NS hull. In the sorbent samples placed directly on the NS hull in addition to ^{137}Cs, ^{134}Cs is recorded; in the sorbent sample placed in the ventilating trunk, ^{54}Mn and ^{60}Co are recorded. In samples of water, taken by bathometers of the deep-sea apparatus *Mir*, higher values of tritium concentrations as compared with background levels were found.

In 1994–1995, in the framework of a marine expedition on the research ship *Academician Mstislav Keldysh*, extended work was carried out in the NS wreckage area to determine the degree of radioactive pollution and radionuclides content of bottom sediments, water, and biota. The work performed can be divided into two categories: that which was executed far from the NS for obtaining data about background values of radionuclide concentrations, and that which was executed near the NS and from possible sources of activity released from reactors and nuclear charges on board the NS.

Serial measurements show that the higher concentration of ^{137}Cs in the seawater of approximately 20 Bq/l has been seen only close to the outlet hole of the ventilation system in the reactor compartment. According to NRB–76/87, the permissible concentration of ^{137}Cs in drinking water is 5.55×10^{2} Bq/l. Thus, the content of this long-lived radionuclide in the seawater is 100 times lower than its permissible concentration in drinking water.

According to 1991–1994 data, the volumetric activity of seawater by $^{239,240}Pu$ in the surface layer in both the background points and in the region of the main work was practically identical. A similar picture was observed for ^{137}Cs and for ^{90}Sr. In 1991–1994, the bottom-dwelling water was characterized by approximately identical values of volumetric activity of $^{239,240}Pu$. The analysis of seawater samples for ^{90}Sr and ^{137}Cs did not show any results that were different from background values.

The radiochemical analysis of bottom sediments samples taken during the 1991–1995 fieldwork shows that the variations of the specific activity of ^{90}Sr in surface layers of bottom sediments (0–2 cm) were between 0.9 and 2.2 Bq/kg.

The results of the analysis to determine the specific activity of ^{239,240}Pu in the surface samples of bottom sediments, taken in close proximity to the NS from the deep-water inhabited apparatus *Mir*, show that between 1991 and 1994 the character of plutonium distribution remained practically unchanged. For example, in surface horizons (0–1 cm) of bottom sediments investigated, the variations of specific activity of plutonium were (0.1–1.6) Bq/kg in 1991, (0.1–1.5) Bq/kg in 1992, (0.1–1.1) Bq/kg in 1993, and (0.04–1.1) Bq/kg in 1994, and are within the variations that are due to global fallouts.

Investigations of the biota have shown that samples taken close to or away from NS do not vary in the content of radionuclides in them.

Identifying the radionuclide source as from the SPR and measurements of specific activity have made it possible to estimate the amount of ^{137}Cs released from the reactor compartment into the seawater. It does not exceed 3.7×10^{11} Bq/yr, which is 30 times less than background content of natural radioactive ^{40}K in 1 km^3 of seawater.

5.3 Characteristics of Available Data

The Ministry of Defense has data on sources and places that produce RWs, the technological operations leading to RW formation, the accumulated waste volumes, their isotopic and chemical composition, specific and volumetric activity, reprocessing technology, and the environmental contamination resulting from the Navy activities.

The information concerning RWs is not classified. Data concerning fresh and SNF for the Navy are classified. An automated database on the problems of RW and nuclear material handling does not exist, but the existing information can be used to create a database. However, at present, the information is not in a uniform format.

Data on the power output of the reactors, radionuclide content in the damaged NS, and the radiation situation in the compartments are still classified.

In three joint reports by the RRC-KI and the Navy, detailed primary information on the scale of the accident in Chazhma Bay and the accompanying radiation situation has been collected. The analysis of this information has allowed conclusions to be drawn about the radiation situation without compromising state secrets or violating RF law.

Data on the character, methods, volumes, and activity of RW dumped by the RF Navy into the Barents Sea, Kara Sea, Sea of Japan, and Sea of Okhotsk from the early 1960s to 1992 have been presented in the report of the Governmental commission on questions connected with marine RW disposal, the so-called *White Book–93*. The Ministry of Defense has evidence of regions where marine RW has been disposed (the radiological and hydrological characteristics of these regions), as well as summary reports on the radioactive effluents deposited into the sea by the Navy. The data on activity of SRWs that sank in the sea are inherently approximate, as the calculation of the activity of SRW dumped was made by a simplified procedure and has been expressed in conventional terms of so-called ^{90}Sr equivalent. However, the information may be considered reasonable for estimating the radioecological situation and predicting its future.

Bibliography

Administration of the President of the Russian Federation, 1993, *Facts and Problems Connected with Radioactive Wastes Dumping in the Seas Surrounding the Territory of the Russian Federation*, Administration of the President of the Russian Federation, Moscow, Russia.

Danilyan, V.A. and Vysotsky, V.L., 1996, Safety issues when handling spent nuclear fuel and radioactive waste in the Pacific Fleet of the Russian Federation, In: *Nuclear Submarine Decommissioning and Related Problems*, L.L. LeSage and A.A. Sarkisov, eds., pp. 201–208, NATO ASI Series, Kluwer Academic Publishers, Dordrecht, Netherlands.

Danilyan, V.A. and Vysotsky, V.L., 1996, Radioecological situation at bases and sites for refueling and recycling of nuclear submarines in the Russian Federation Pacific Fleet, In: *Nuclear Submarine Decommissioning and Related Problems*, L.L. LeSage and A.A. Sarkisov, eds., pp. 321–330, NATO ASI Series, Kluwer Academic Publishers, Dordrecht, Netherlands.

Gladkov, G.A. and Sivintsev, Yu.V., 1994, The radiation situation in the area where the nuclear submarine *Komsomolets* sunk, *Atomic Energy*, **77**(5):379–386.

Joint Russian/Norwegian Expert Group for Investigation of Radioactive Contamination in Northern Areas, 1996, *Radioactive Contamination at Dumping Sites for Nuclear Waste in the Kara Sea. Results from the Russian–Norwegian 1993 Expedition to the Kara Sea*, Joint Russian/Norwegian Expert Group for Investigation of Radioactive Contamination in Northern Areas, Oslo, Norway.

Kazennov, A.Yu., 1995, Monitoring of sea contamination, In: *Report from the First Aleksandrov Memorial Meeting*, pp. 14–25, Russian Research Center-Kurchatov Institute (RRC-KI), Moscow, Russia.

Kiknadze, O.E. and Sivintsev, Yu.V., 1995, Radioecological danger of long-lived radionuclides in a nuclear reactor dumped in the Arctic, In: *Environmental Radioactivity in the Arctic*, H. Strand and A. Cooke, eds., pp. 152–155, Scientific Committee on the Environmental Radioactivity in the Arctic, Osteras, Norway.

Kiknadze, O.E. and Sivintsev, Yu.V., 1995, Radioecological danger of long-lived radionuclides in a nuclear reactor dumped in the Arctic, *Atomic Energy*, **79**(3):204–211.

Kuznetsov, Yu.V. and Legin, V.K., 1993, The radioactive contamination of the northern seas: Approaches to the assessment of the impact on the marine environment, In: *Radioactivity and Environmental Security in the Oceans*, pp. 379–394, Woods Hole Oceanographic Institution, Woods Hole, MA, USA.

Kuznetsov, Yu.V., Nosov, N.A., and V.K. Legin, 1993, Dumping of radioactive waste in the sea environment: Scientific and practical aspects, In: *Environmental Radioactivity in Arctic and Antarctic*, H. Strand and E. Holm, eds., pp. 37–53, The Scientific Committee on the Environmental Radioactivity in the Arctic and Antarctic, Osteras, Norway.

Kuznetsov, A.P. *et al.*, 1996, Radionuclides in the benthofauna in the region of the location of the nuclear submarine *Komsomolets*, *Biology Bulletin*, **23**(4):387–390.

Petrov, O.I., 1993, Environmental and sanitation problems related to radioactive waste dumping into the Arctic Seas, In: *Environmental Radioactivity in Arctic and Antarctic*, H. Strand and E. Holm, eds., pp. 97–102, The Scientific Committee on the Environmental Radioactivity in the Arctic and Antarctic, Osteras, Norway.

Petrov, O.I., 1993, Radioactive wastes generated by ship nuclear power plants, In: *Assessment of Actual and Potential Consequences of Dumping of Radioactive Wastes into Arctic Seas*, pp. 71–74, International Atomic Energy Agency (IAEA), Vienna, Austria.

SDBME (Special Design Bureau of Marine Engineering) Rubin, 1995, Development of plans and specifications for utilizing operating, damaged, and newly built nuclear submarines, Report presented at the *Scientific Practical Conference*, 15–16 March, Severodvinsk, Russia.

Sivintsev, Yu.V., 1993, Monitoring of the radioactive contamination near sunken submarines, In: *Assessment of Actual and Potential Consequences of Dumping of Radioactive Wastes into Arctic Seas*, pp. 75–80, International Atomic Energy Agency (IAEA), Vienna, Austria.

Sivintsev, Yu.V., Vysotsky, V.L., and Danilyan, V.A., 1994, Ecological consequences of the radiation accident on the nuclear submarine in the Chazhma Bay, *Atomic Energy*, **76**(2):158–160.

Vinogradov, M.E., 1996, *Oceanologic Research and Submarine Technical Works at the Place of NS Komsomolets Wreck*, Nauka, Moscow, Russia.

Chapter 6

Nuclear Research Reactors and Nuclear Research Centers

6.1 Introduction

The nuclear research centers of the former Soviet Union (FSU) conduct studies in the fields of nuclear physics, solid-state physics, radiation material science, and radioisotope production. Most were constructed in the 1940s and 1950s, and the main facilities were put into operation in the late 1960s. As a rule, these centers were complex systems – that is, they involved not only the research reactors but also hot cells for material tests, research laboratories, radioactive waste (RW) handling facilities, etc. The research reactors installed in the nuclear centers are very diverse in their designs, power levels, material used, and operation mode. These reactors belong to many different departments. A brief summary of the research reactors is given in *Table 6.1* at the end of this chapter.

6.2 Characteristics of Accumulated Radioactive Materials and Radioactive Wastes

The information in this section was the most current available as of 1 July 1995.

6.2.1 Russian Research Center – Kurchatov Institute (RRC–KI), Moscow, Russia

Reactor F–1

This reactor contains 46.411 kg of natural uranium in the form of cylindrical metallic slugs, balls, and pellets of UO_2 and U_3O_8, as well as about 41 kg of 2% enriched uranium cylindrical metallic slugs. The burnup in the natural uranium slugs is 6 kW-days (kWd) per slug; in the 2% enriched uranium slugs it is 2 kWd/slug. Total activity is about 6×10^{11} Bq.

The reactor contains no RW, and no radionuclides have been discharged to the environment. The ^{41}Ar release is negligible because of the low thermal power of the reactor. The reactor has no radioactive material (RM) or RW storages. The 430-ton graphite stack is the only potential RW if the reactor were to go out of service, and then the nuclear fuel would need to be removed.

Reactor VVR–2

The reactor was decommissioned in 1983. The spent fuel that accumulated in the reactor from 1956 to 1982 is kept in a cooling pond, which includes 2,657 fuel rods with an initial enrichment of 10% and 1,447 fuel rods with an initial enrichment of 36%. The total mass of fissile materials (FMs) is 241.6 kg. The total activity of fission products (FPs) in the spent fuel is about 2×10^{15} Bq. The uranium activity is 5.2×10^{9} Bq. No radionuclides have been discharged to the environment. The ^{41}Ar released to the environment did not exceed 0.5 Ci/hr. The reactor has a liquid radioactive waste (LRW) storage facility consisting of two metallic tanks with a capacity of 50 m^3 each. They contain 14 m^3 of low-active LRW with a total activity of 2.6×10^{10} Bq. The storage facility is over 30 years old.

For solid radioactive wastes (SRWs), two deep steel tanks with a total volume of 87 m^3 contain about 70 tons of highly active SRW (components of the dismantled reactor). The total highly active SRW activity is 3.1×10^{11} Bq.

Reactor RFT

The reactor was put out of service in 1962. The spent fuel of the reactor that was accumulated during the period of its operation from 1953 to

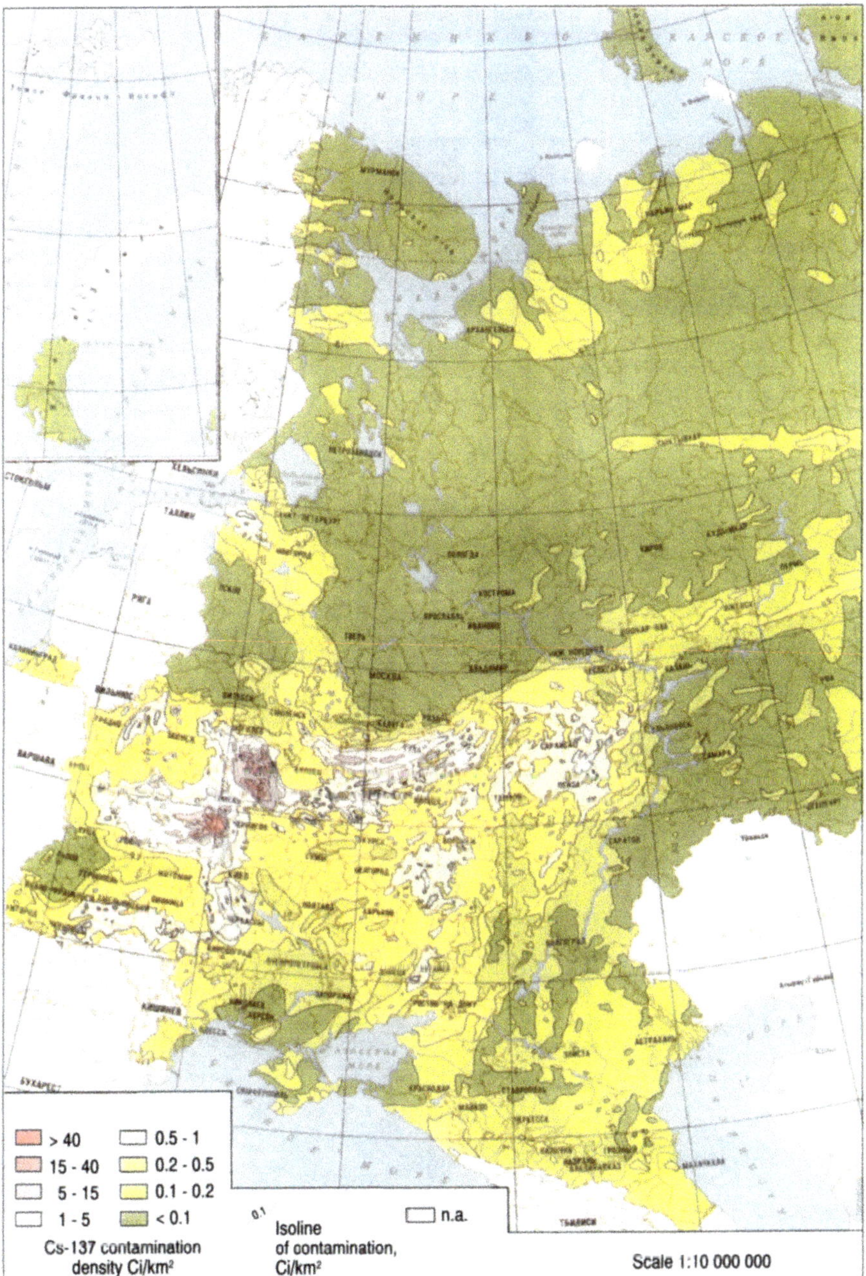

Color Plate 1. ^{137}Cs contamination densities of Belorussia, Ukraine, and the European part of Russia.

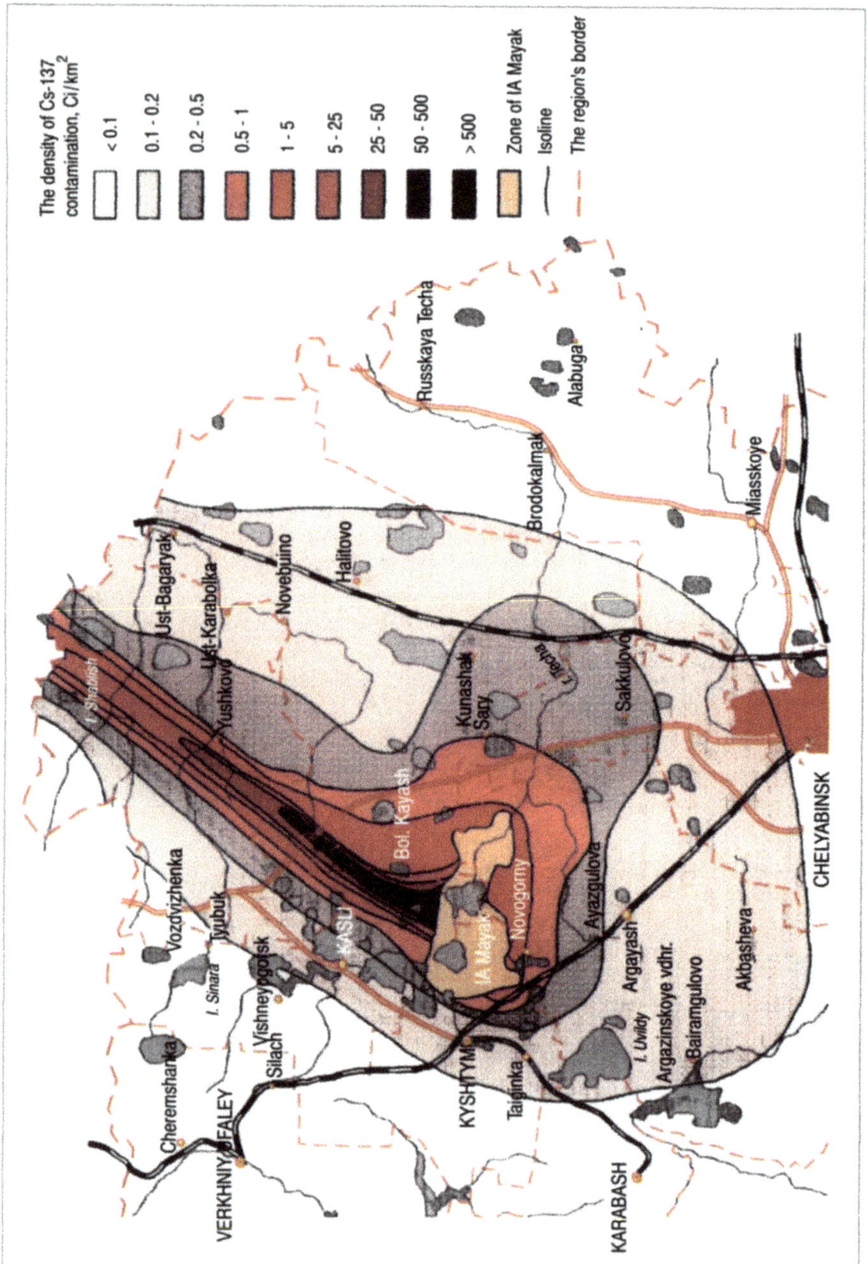

Color Plate 2. ^{90}Sr contamination of the northern part of the Chelyabinsk region.

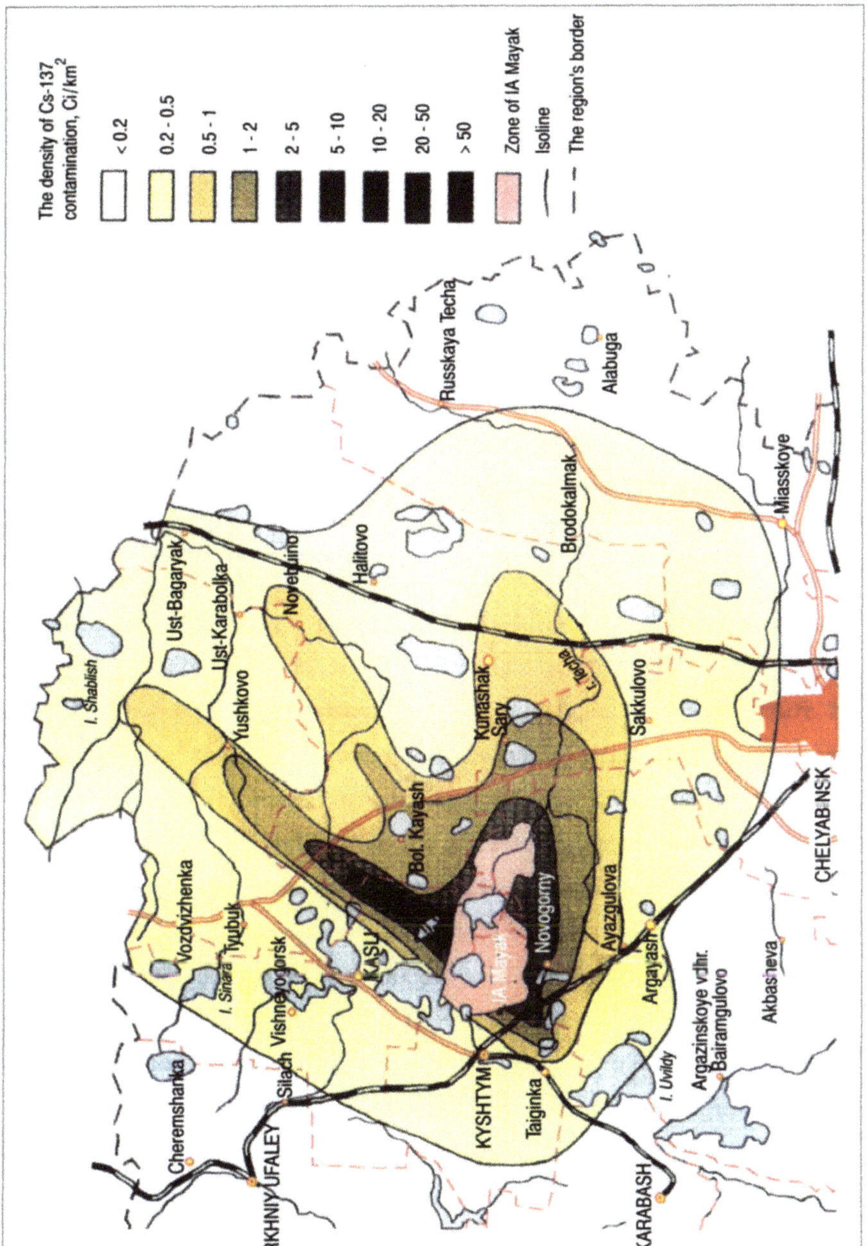

Color Plate 3. ^{137}Cs contamination of the northern part of the Chelyabinsk region.

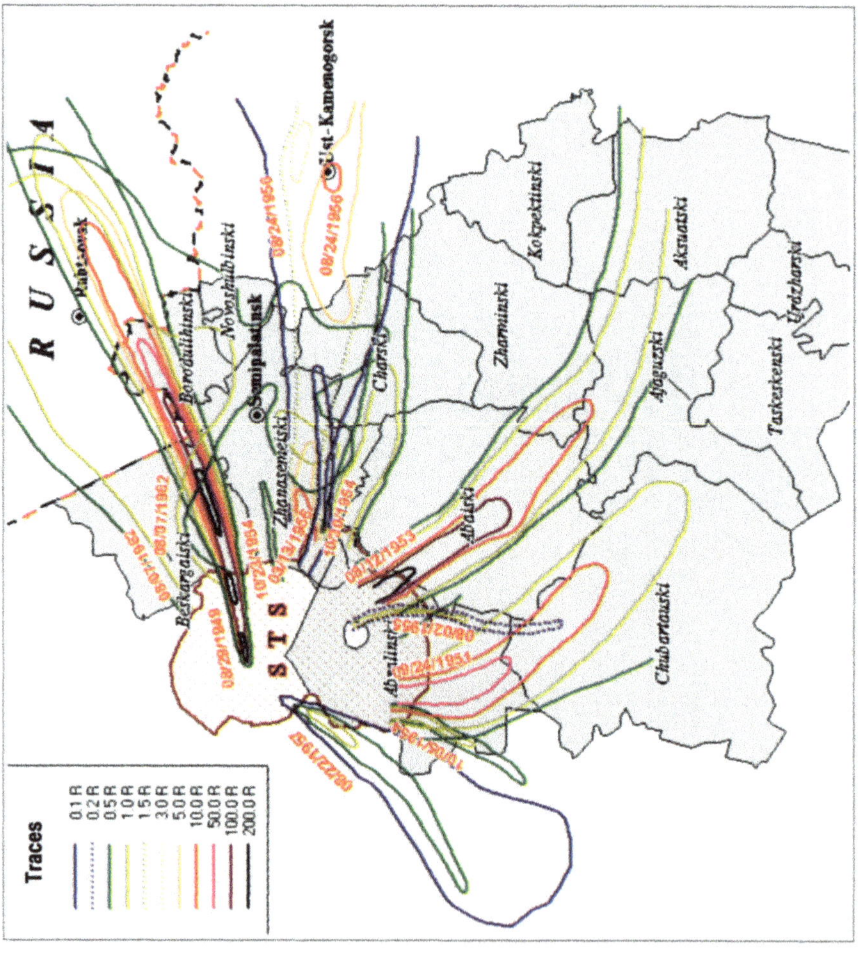

Color Plate 4. The Semipalatinsk test site and axes of the most significant dose-generating local traces of radioactive contamination.

1958 is kept in the dry storage of the MR reactor (initial enrichment is 10%). The FM weight totals 302.44 kg. The total activity of FPs in the spent fuel is about 3×10^{15} Bq; uranium activity is about 5×10^{9} Bq. No radionuclides have been discharged to the environment. The steel reactor vessel with the graphite stack remains in the reactor building. Potential RWs from the dismantling of the reactor vessel include the steel vessel parts, graphite, and part of the concrete of the shielding well housing the vessel. Therefore, the ultimate quantity of RWs and their activity cannot be determined until after the dismantling is completed.

Reactor MR

The reactor was shut down in 1993. The first decommissioning phase has been completed; the fuel was discharged and, after being held in the storage pool, put into the MR dry storage facility. The subsequent decommissioning phases have been undertaken.

The spent fuel that accumulated in the reactor from 1986 to 1993 – 187 fuel assemblies (FAs) with an initial enrichment of 90% – is kept in the dry storage facility of the MR reactor. The FM mass totals 63.3 kg. The total activity of FPs in the spent fuel is about 4.4×10^{16} Bq and the uranium activity is about 4.4×10^{9} Bq. About 200 spent loop assemblies are in storage, and their FP activity is about 5×10^{16} Bq.

The dry storage facility is located in an isolated room of a building near the reactor, and has three sections. Section I is intended for storage of FAs of the RFT reactor (rows 1–16) and pilot loop FAs (rows 17–24). The number of cells in Section I totals 192. The spent FAs of the RFT reactor are kept in the cells in aluminum-sealed casks. The spent loop FAs are put into the cells in sealed stainless steel casks.

Section II is used to store spent FAs from the MR reactor, which are transferred there after being held in the storage pool in the reactor building. The number of Section II cells totals 216. In this section, the MR FAs can be stored in two rows.

Section III is intended for storage of the IRT–3M-type FAs in aluminum casks. There are 117 cells in this section. The sections are covered from the top with a 46-cm-thick protective steel plate with charging openings that are provided with plugs. The storage facility age is about 40 years. No radionuclides have been released to the environment.

The reactor has three 30-m^3 metallic tanks for storage of low-activity LRW. They contain 20 m^3 of liquid wastes with a total activity of

7.4×10^8 Bq. There is a 30-m^3 metallic tank for medium-activity LRW containing 20 m^3 of wastes with a total activity of 7.4×10^{10} Bq. The storage facility age is about 40 years.

For storage of SRWs, deep concrete compartments and a concrete storage facility with 127 cells are provided. They contain SRW with a total activity of about 1,015 Bq. A considerable number of spent loop channels are in the storage pool. Some sections of these channels have a high-specific activity.

The amount of SRW formed while putting the reactor and its 10-loop facilities out of service can only be estimated after these operations are completed.

Reactor IRT

The reactor was put out of service in 1979, and it was replaced by the IR–8 reactor. The dismantled in-pool equipment (reactor vessel, ejector, sections of the horizontal experimental channels) was transferred to the storage facility at Moscow Industrial Association (MNPO) Radon. The spent fuel was sent to the Industrial Association (IA) Mayak for reprocessing.

Reactor IR–8

There are 16 FAs (initial fuel enrichment is 90%; the uranium content in each FA is 0.294 kg of FM) in the core. The burnup averaged over the core volume is about 25%. The total activity of FPs in the core, with the reactor in operation, is 5.7×10^{16} Bq; uranium activity is 2.6×10^8 Bq.

The pool, designed for storage of up to 120 spent FAs, contains 36 FAs that have been discharged out of the reactor between 1989 and 1995. The mass of uranium in the FAs totals 5.79 kg of FM. The activity of FPs in 36 FAs totals 3×10^{15} Bq; the uranium activity is 3.9×10^8 Bq. The reactor has two underground steel tanks each capable of holding 50 m^3 of LRW. There are 20 m^3 of low-activity LRW, with a total activity of 7.4×10^6 Bq. The storage facility age is about 40 years. No radionuclides have been released to the environment. The ^{41}Ar released to the environment did not exceed 1 Ci/hr. The reactor has no storage for SRWs.

Reactor IIN–3M Gidra

The Gidra is a homogeneous reactor with a core holding 40 liters of aqueous solution of UO_2SO_4. It contains 3.44 kg of uranium enriched to 90%. The burnup is insignificant. The activity totals about 3×10^8 Bq. No radionuclides have been released to the environment. Practically no ^{41}Ar has been released to the environment because of a low power and pulse mode of operation. The reactor has neither fuel nor RW storage facilities.

Reactor GAMMA

The core of the GAMMA has 69 FAs, each containing 0.56 kg of FM with the initial enrichment higher than 20%. The average burnup in the FA is 70 kWd. The fuel activity totals about 1.4×10^9 Bq.

The reactor has a 10-m^3 metallic container for LRWs, containing 5 m^3 of low-active LRW. The activity of the liquid wastes totals 2.0×10^5 Bq. The storage facility age is 13 years. No radionuclides have been released to the environment. The ^{41}Ar released to the environment is insignificant. There are no SRWs in the reactor.

Reactor Argus

The Argus is a homogeneous reactor with a core containing 22 liters of aqueous solution of UO_2SO_4. It contains 1.71 kg of uranium enriched to 90%. The burnup is insignificant. The fuel activity totals 1.2×10^8 Bq. No radionuclides have been released to the environment. The ^{41}Ar released to the environment is insignificant because of the low level of the thermal power. The reactor has neither fuel nor RW storage facilities.

Reactor OR

The core of the OR has 375 fuel rods, each containing 0.025 kg of uranium of 36% initial enrichment. The average burnup in the fuel rod is 200 kWd. The activity of FPs in the core, with the reactor in operation, totals about 3×10^{16} Bq; the uranium activity is about 3.3×10^8 Bq. For storing RW produced in the reactor operation, the VVR–2 reactor storage facilities are used.

No radionuclides have been released to the environment. The ^{41}Ar released to the environment has not exceeded 0.05 Ci/hr.

Reactor Romashka

The Romashka was shut down in 1966 and has been decommissioned. The spent fuel accumulated during the reactor's operation from 1964 to 1966 (44.5 kg of FM) is in dry storage. The activity of the FPs in the spent fuel totals 2.9×10^{12}, and the uranium activity is 3.2×10^{9} Bq.

No radionuclides have been released to the environment. The ^{41}Ar released to the environment was insignificant.

Reactor Topaz–2

The Topaz–2 was shut down and put out of service in 1986. The spent fuel accumulated during operations from 1973 to 1986 is kept in dry storage (five core loads totaling 112.3 kg of FMs). The activity of FPs in the spent fuel totals 4.2×10^{13} Bq, and the uranium activity is 8.5×10^{10} Bq. No radionuclides have been released to the environment. The ^{41}Ar released to the environment was insignificant.

Critical Test Facilities SF–1, SF–3, SF–5, SF–7, Kvant, Physical Model of the MR Reactor, Delta, UG, RBMK, V–1000, ASTRA, EFIR–2M, Mayak, NARCISS M2, GROG, P, and ISKRA

The fuel used in the critical test facilities (CTFs) can be considered practically fresh since the power of most CTFs does not exceed 100 W, and they do not operate in continuous mode. Both natural uranium and fuel enriched from 1.6% to 96% is used in the CTFs. The quantity of fuel contained in all CTFs totals about 48,270 kg. Its total activity is about 7×10^{11} Bq.

No radionuclides have been released from the CTFs to the environment. No ^{41}Ar has been released to the environment because of the low CTF power level.

At the RRC-KI's CTFs, there have been no accidents that could have resulted in environmental RW contamination, including in the sanitary-protection zone (SPZ).

6.2.2 RF State Research Center – Institute of Theoretical and Experimental Physics, Moscow, Russia

Reactor TVR

The TVR was shut down in 1986 and is now in the decommissioning phase. The spent fuel accumulated during the reactor operations from 1949 to 1986 was transferred to IA Mayak. No radionuclides have been released to the environment. The heavy water moderator-coolant contaminated with tritium remains in the reactor hall. No LRW is in the reactor.

The low-active SRWs are stored in the concrete biological shielding of the reactor. These wastes have a mass of 995 tons, a volume of 414 m^3, and an activity of 1.87×10^{10} Bq. Medium-active SRWs are also stored in the concrete biological shielding of the reactor; they have a mass of 84.14 tons, a volume of 33.14 m^3, and an activity of 215×10^{10} Bq. High-active SRWs are kept in a special shielded well; they have a mass of 1.1 tons, and an activity of 12.54×10^{10} Bq. This activity is primarily due to ^{60}Co and ^{137}Cs.

Because RWs are included in parts of the steel vessel, heavy water, and the reactor well concrete, RWs would be released if the reactor were dismantled. The quantity of RWs and their activity can be determined only after the dismantling is completed.

Critical Test Facility Mayak

The core of the Mayak Test Facility contains 300 FAs with an initial enrichment of 90%. No burnup has occurred. The CTF has neither RW nor fuel storage facilities. No radionuclides have been released to the environment, and no ^{41}Ar has been released to the environment.

6.2.3 Moscow State Institute of Engineering Physics (Engineering University), Moscow, Russia

Reactor IRT

The core of the IRT contains 15 or 16 FAs with an initial fuel enrichment of 90%. Uranium content in the six-tube FAs is 0.294 kg, and in the eight-tube FAs it is 0.333 kg FM. The ^{235}U burnup average over the core is about 20%. The activity of FPs in the core of the operating reactor totals 2.5×10^{17} Bq, and the uranium activity is 3.6×10^{8} Bq. The spent fuel discharged out of the reactor up to 1982 was transferred to IA Mayak. The spent fuel discharged from the reactor between 1982 to 1991 is stored in a reactor storage well (an aluminum tank located near the reactor pool in the concrete shielding). The number of FAs kept in the storage well totals 52 and the total weight of uranium in the FAs is 19 kg of FM. The activity of FPs in the fuel totals 0.3×10^{17} Bq, and the uranium activity is 7×10^{8} Bq.

The reactor is provided with two 100-m^3 steel reservoirs, which contain 30 m^3 of low-active LRW (the activity totals 107 Bq). The storage facility is 30 years old. Low-active solid wastes are temporarily kept in paper and polyethylene bags.

No radionuclides have been released to the environment, and the ^{41}Ar released to the environment did not exceed 0.3 Ci/hr. For the period of the reactor operations, there were no accidents with RW releases that could have contaminated the environment, including the SPZ. Potential RWs that may be released in case of the reactor dismantling include in-pool components of the reactor, a part of the biological shielding, etc.

6.2.4 Research and Design Institute of Power Engineering, Moscow, Russia

Reactor IR–50

The IR–50 was shut down for restoration in 1993. The spent fuel that accumulated in the reactor during its operations from 1961 to 1993 – 603 FAs with an initial enrichment of 10% – is stored in a wet storage tank. The fuel burnup is about 1%. The total activity of FPs in the spent fuel is 0.25×10^{15} Bq. No LRW is in the reactor.

An insignificant amount of SRW contained in the reactor is gradually being transferred to MNPO Radon. No radionuclides have been released to the environment, and the ^{41}Ar released to the environment was insignificant. During the reactor operations there were no accidents that could have resulted in the RW contamination of the environment, including the SPZ.

6.2.5 Research Instrumentation Institute, Litkarino, Moscow Region, Russia

Reactor IRV

The IRV was shut down in 1990 for restoration.

The reactor storage facility contains 22 FAs with an initial enrichment of 90%, which were accumulated during the reactor's operation from 1974 to 1990. The burnup of the fuel is 10.35 MWd/FA. The ^{235}U in the spent FAs totals 3.46 kg. The total FP activity in the fuel is 7.4×10^{12} Bq.

The low-active liquid wastes are stored in sumps and tanks of the special water treating station; their volume is 5 m^3 and their activity is 1.8×10^6 Bq. The low-active solid wastes are stored on special shelves and in casks; their mass is 0.165 tons, the volume is 1 m^3, and the activity is 1.2×10^7 Bq. The medium-active SRWs are kept in special containers; their mass is 30.76 tons, the volume is 40 m^3, and the activity is 3×10^{12} Bq.

No radionuclides have been released to the environment, and the ^{41}Ar released to the environment was insignificant. During the reactor operation, there were no accidents that could have led to RW contamination of the environment, including the SPZ.

6.2.6 RF State Research Center – Konstantinov Institute of Nuclear Physics, Gatchina, Leningrad Region, Russia

Reactor VVR–M

The core of the VVR–M contains 145 FAs with an initial fuel enrichment of 90%. The uranium content per FA is 0.074 kg. The activity of FPs in the core totals 1.8×10^{18} Bq, and the uranium activity is 7×10^8 Bq.

The spent fuel that was discharged out of the reactor by 1980 was transferred to IA Mayak. The spent fuel that was discharged out of the reactor from 1980 to 1995 is kept in the storage pool near the massive biological reactor shielding below the floor of the experimental hall. The number of FAs in the storage facility totals 802. The total mass of uranium in FA is 43 kg. The total activity of FPs is 2×10^{17} Bq, and the uranium activity is 3×10^9 Bq.

The reactor has two underground tanks and one surface steel tank, each holding 300 m^3. The tanks contain 150 m^3 of low-active LRW with a total activity of 5.5×10^9 Bq.

The storage facility age is about 36 years. The reactor has no permanent SRW storage; instead, low-active SRWs are temporarily kept in bags and casks.

No radionuclides have been released to the environment, and the ^{41}Ar released to the environment did not exceed 2 Ci/hr. During reactor operations, there were no accidents that could have resulted in the RW contamination of the environment, including the SPZ.

Critical Test Facility with the Physical Model of PIK Reactor

The core of the PIK physical model contains 18 FAs with 30.6 kg of uranium enriched with ^{235}U to 90%. The uranium activity totals about $2{\times}10^9$ Bq. No radionuclides have been released to the environment.

6.2.7 RF State Research Center – Research Institute of Atomic Reactors, Dimitrovgrad, Ulyanovsk Region, Russia

Reactor SM–3

The core of the SM–3 contains 28 FAs with an initial fuel enrichment of 90%. The uranium content in a full FA is 1.05 kg; the uranium content in the FA for target irradiation is 0.89 kg. The activity of FPs in the core totals $16.8{\times}10^{18}$ Bq. Two interim repositories of spent FA are mounted in the reactor vessel. They can accommodate up to 16 FAs without depressurizing the reactor. The system of temporary FA storage in the reactor involves three cooling ponds made of stainless steel. The cooling ponds of the storage facility can contain up 852 FAs. The age of the storage facility is about 35 years.

For collection and interim storage of LRWs, there are eight stainless steel tanks and two carbon steel tanks, each holding 3 m^3 of LRW. For collection of high-specific activity SRWs, a reinfored compartment is used; slightly active solid wastes are collected in a container. No radionuclides have been released to the environment.

Reactor MIR–M1

The core of the MIR-M1 contains 48 FAs with an initial fuel enrichment of 90%. Uranium content in each FA is 0.39 kg, and the activity of FPs in the core totals $6.8{\times}10^{18}$ Bq.

The spent FAs discharged out of the reactor between 1966 and 1996 are stored in the cooling pond made of stainless steel. The cooling pond can contain up to 356 FAs and up to 22 experimental tubes. The storage facility age is about 30 years. No radionuclides have been released to the environment.

Reactor BOR–60

There are 100 FAs with an initial fuel enrichment of 90% in the core of the BOR–60. The activity of FPs in the core totals 10.6×10^{18} Bq, and the uranium activity is 7.8×10^{8} Bq. The spent FAs discharged from the reactor between 1969 and 1996 are stored in the stainless steel cooling pond. The cooling pond can contain up to 356 FAs and up to 22 experimental tubes. The storage facility age is about 28 years. No radionuclides have been released to the environment.

Reactor RBT–6

The core of the RBT–6 contains 56 spent FAs from the SM–3 reactor with an initial fuel enrichment of 63%. The activity of FPs in the core totals 1.1×10^{18} Bq. There is no special spent fuel storage in the reactor, as it uses the spent FAs of the SM–3 reactor. No radionuclides have been released to the environment.

Reactor RBT 10/1

The core of the RBT–10/1 contains 78 spent FAs from the SM–3 reactor with an initial fuel enrichment of 63%. The activity of FPs in the core totals 1.8×19^{18} Bq. The reactor has no special spent fuel storage, as it uses the spent FAs of the SM–3 reactor. No radionuclides have been released to the environment.

Reactor RBT–10/2

There are 78 FAs with an initial fuel enrichment of 63% in the core of the RBT–10/2 reactor. The activity of FPs in the core totals 1.8×10^{18} Bq. The reactor has no special spent fuel storage, as it uses the spent FAs of the SM–3 reactor. No radionuclides have been released to the environment.

Reactor VK-50

The core of the VK–50 contains 72 FAs with an initial fuel enrichment of 2%, 2.4%, or 3%. The activity of FPs in the cores totals 35.2×10^{18} Bq. LRWs are stored in a 50-m^3 stainless steel tank.

Critical Test Facilities with the Physical Models of the SM–3 and MIR–M1 Reactors

The fuel used in the CTFs are considered practically fresh, as the power of most CTFs does not exceed 100 W and they do not operate in continuous mode. The CTF uses 90% enrichment fuel. The activity of the fuel totals about 10^{11} Bq.

No radionuclides have been released to the environment, and no ^{41}Ar has been released to the environment because of low CTF power.

The RF State Research Center–Research Institute of Atomic Reactors (RF SRC–NIIAR) has a centralized system of storing spent FAs as well as a system for collection, storage, and disposal of LRWs and SRWs from all the institute's installations. The spent FA storage consists of two 7-m-deep pools containing 5,200 FAs with 85 tons of uranium, including 4 tons of uranium of 90% enrichment.

The shares of waste from the reactors, radiochemical production, and special laundry are 60%, 10%, and 15%, respectively, with the remaining 15% unspecified. Low- and medium-active LRWs are transferred by special sewage systems to a 200-m^3 tank and four reservoirs, each holding 2,000 m^3. The activity of these wastes totals 1.5×10^{12} Bq. The waste activity is determined by isotopes ^{134}Cs, ^{137}Cs, ^{90}Sr, ^{131}I, ^{103}Ru, ^{106}Ru, ^{60}Co, etc. After mixing, settling, and neutralizing, the low-active LRWs are pumped into the underground repository of a semi-industrial test site, which is a system of injection and observation wells provided with appropriate process equipment and systems of technical control and radiation monitoring. The activity of RWs pumped into the ground repository is 4.1×10^{15} Bq.

The high-active LRWs are transferred by the special sewage system to seven reservoirs: Three have a capacity of 500 m^3, two have a 600-m^3 capacity, and two are capable of holding 6,000 m^3. The activity of these wastes totals 9.3×10^{14} Bq, and the volume is 3,300 m^3. The activity of the wastes is determined by isotopes ^{134}Cs, ^{137}Cs, ^{90}Sr, ^{131}I, ^{103}Ru, ^{106}Ru, ^{60}Co, etc. The low-active SRWs are stored in paper bags and transferred to trenches that, upon being filled, are covered with a 1.5–2.0-m-thick layer of soil. Around the bunker, a system of wells is provided to control groundwaters. The activity of these annually produced wastes totals 7.4×10^{11} Bq. The high-activity SRWs are shipped by special vehicles to high-active waste burial facilities in reinforced concrete bunkers. The activity of these wastes totals 3.3×10^{16} Bq, and volume is in excess of 1,700 m^3.

No accidents have occurred with the CTF or reactors of the RF SRC–NIIAR that would have led to RW contamination of the environment, including the SPZ.

6.2.8 Research Institute of Nuclear Physics of Tomsk Polytechnical University, Tomsk, Russia

Reactor IRT–T

The core of the IRT–T contains 20 FAs with an initial fuel enrichment of 90%. The uranium content in a six-tube FA is 0.294 kg and in an eight-tube FA it is 0.333 kg. The activity of FPs in the core of the operating reactor totals 6×10^{17} Bq. The uranium activity is 2.2×10^{8} Bq. The spent fuel discharged out of the reactor is stored in a storage well. The number of FAs in the storage totals 64, and the total mass of uranium is about 34 kg. The total activity of FPs is 6×10^{16} Bq, and uranium activity is 9×10^{8} Bq.

The reactor is provided with three steel tanks for LRW storage: Two have a capacity of 100 m^3 each, and one has a capacity of 325 m^3.

The total quantity of low-active LRW contained in these tanks is 100 m^3, 25 m^3, and 50 m^3, in the first, second, and third tanks, respectively. The total activity of LRW is 3×10^{10} Bq, and includes ^{60}Co (4.8×10^{8} Bq), ^{134}Cs (10^{8} Bq), and ^{137}Cs (3.8×10^{8} Bq). The storage facility is about 30 years old. The reactor has no SRW storage facilities. No radionuclides have been released to the environment, and the ^{41}Ar released to the environment did not exceed 1 Ci/hr.

During the reactor's operation, there were no accidents that could have resulted in the RW contamination of the environment including the SPZ.

6.2.9 Sverdlovsk Branch of Research and Design – Institute of Power Engineering, Zarechny, Sverdlov Region, Russia

Reactor IVV–2M

The core of the IVV–2M contains 42 FAs with an initial fuel enrichment of 90%. The uranium content in each FA is 0.25 kg. The activity of FPs in the core of the operating reactor totals 2.8×10^{18} Bq. The spent fuel discharged from the reactor by 1991 was transferred to IA Mayak. The spent fuel discharged from the reactor between 1992 and 1996 is in a regular storage pool. Low-active LRWs are stored in special 100-m^3 tanks. Their total activity is 4.4×10^{11} Bq. The low-level SRWs are stored in special containers at the work site; their annual buildup is 35 m^3. Medium-active SRWs are kept in special containers at the work site; their annual buildup is 15 m^3. The low- and medium-active SRWs are transferred for disposal to the regional division of the combine Radon (Pyshma), and high-active RWs are transferred to the IA Mayak.

6.2.10 RF State Research Center – Institute of Physics and Power Engineering (RF SRC–FEI), Obninsk, Kaluga Region, Russia

The RF SRC–FEI, set up in 1946, has two research reactors and a large number of critical assemblies. A large quantity of various RWs has accumulated.

Reactor AM

The core of the AM contains 122 FAs, of which 31 FAs have an initial enrichment of 4.4% and 91 FAs have an initial enrichment of 10%. The FM content in the FAs is 120 kg. The fuel burnup in the FA totals 60 MWd/FA and 100 MWd/FA, respectively. The total fuel activity is 2×10^{18} Bq.

The spent fuel discharged from the reactor for the period from 1962 to 1996, is stored in the cooling pond. The number of FAs in the cooling pond totals 297. The total mass of FM in the fuel is 1,250 kg. In the centralized SNF storage facility of the RF SRC–FEI are 4,274 fuel rods of reactor AM, which had been discharged from 1962 to 1996 with a burnup of 25–35 MWd/FR. The reactor has no LRW storage facility. The quantity of SRW totals 90 tons, its volume is 30 m^3, and its activity is

5.55×10^{12} Bq. The wastes are stored in a 148-m^3 concrete bunker. No radionuclides have been released to the environment.

Reactor BR–10

The core of the BR–10 contains 97 FAs with an initial enrichment of 90%. The total mass of FM is 130 kg. The fuel burnup in the core is 5,500 MWd/t, and the total activity of FPs is 1.6×10^{18} Bq. The spent fuel discharged from the reactor between 1959 and 1991 is placed in the centralized SNF storage facility of RF SRC–FEI. The number of FAs in the storage facility totals 497, and the fuel burnup is 20–35 MWd/FA. The total quantity of FMs in the FAs is 0.93 tons. The total activity of FPs is 3.7×10^{16} Bq.

The LRWs are stored in a metallic tank. Their volume totals 14.6 m^3, and their activity is 1.33×10^{13} Bq. The high-active SRWs are stored in a concrete bunker; their mass is 4.5 tons, their volume is 0.9 m^3, and the total activity is 6×10^{13} Bq. No radionuclides have been released to the environment.

Critical Test Facilities BR–1, BFS–1, RF–GS, MATR–2, PS–2, T–2, SGO, Strela, BFS–2, KOBR, AMBF–2, FS–1M, K–1

The fuel used in the CTFs is nearly fresh, as the power of most CTF does not exceed 100 W and they do not operate in continuous mode. Both natural uranium and fuel of 1.6% to 96% enrichment are used in the CTFs. The content of FMs in spent FAs of RF SRC–FEI totals 14.52 tons. The total activity of FPs is 4×10^{17} Bq. No radionuclides have been released to the environment.

6.2.11 Branch of RF State Research Center – Karpov Institute of Physical Chemistry, Obninsk, Kaluga Region, Russia

Reactor VVR–C

The core of the VVR–C contains 70 FAs with an initial fuel enrichment of 36%. The uranium content in a FA is 0.303 kg. The total activity of FPs in the core of the operating reactor is 1.4×10^{18} Bq, and the uranium activity is 6.5×10^{8} Bq. Spent fuel discharged from the reactor before 1991 was transferred to IA Mayak. The 242 FAs discharged from the reactor between 1991 and 1995 are kept in the storage pool. The mass of uranium in the FA totals about 63 kg. The total activity of FPs in them is 2×10^{17} Bq, and the uranium activity is 1.8×10^{9} Bq.

The reactor has a 200-m^3 stainless steel tank for the storing of low-active LRWs. It contains 14 m^3 of wastes with a total activity of 2.6×10^{10} Bq. The storage facility age is about 31 years. For storing medium-active LRW, a 200-m^3 stainless steel tank, divided into two compartments, is provided. It now contains no wastes.

There are two storage facilities for SRWs: a 150-m^3 wood-lined deep cavity in the concrete floor for low-active SRW, where 60 m^3 of wastes are contained; and a 225-m^3 reinforced concrete semideep cavity for high-active SRW, where 70 m^3 of wastes with a total activity of about 3×10^{11} Bq are stored. The storage facility age is about 31 years.

No radionuclides have been released to the environment, and the ^{41}Ar released to the environment does not exceed 7.4×10^{10} Bq/hr. During reactor operations no accidents have occurred. The environment, including the SPZ, has not been contaminated with RW.

6.2.12 Latvia AS Institute of Physics, Riga District

Reactor IRT–M

The core of the IRT–M contains 21 FAs with an initial enrichment of 90%. The uranium content in the FAs is 0.294 kg and 0.333 kg in the six- and eight-tube FAs, respectively. The activity of FPs in the core of the operating reactor totals 5×10^{17} Bq. The uranium activity is 3.8×10^{8} Bq. Spent fuel discharged from the reactor before 1987 was transferred to IA Mayak. Spent fuel discharged from the reactor between 1987 and 1994 is stored in a storage well. The total weight of uranium in the 44 FAs within the storage well is 8.8 kg, with a total activity of FPs estimated at 2×10^{16} Bq, and the uranium activity is 6×10^{8} Bq.

The reactor is provided with two 100-m^3 underground metallic tanks for storing LRWs. No data are available about the content of low-active LRWs in these tanks. The age of the storage facility is about 34 years. The reactor has no storage facility for SRWs.

No radionuclides have been released to the environment, and the ^{41}Ar released to the environment did not exceed 1.85×10^{10} Bq/hr. No accidents have occurred during reactor operations. The environment, including the SPZ, has not been contaminated with RW.

6.2.13 Belorussia AS Nuclear Power Institute, Sosny, Minsk District

Reactor IRT–M

The IRT–M was shut down and put out of operation in 1988.

Practically all of the accumulated spent fuel was transferred to IA Mayak. However, 24 EK–10 fuel rods that were discharged in the 1970s remain in the reactor storage pool. These rods have an initial fuel enrichment of 10%, and the ^{235}U content in the fuel rod is 8 g. The fuel burnup is 20%. The total activity of FPs in the FAs is estimated at about 10^{14} Bq.

No radionuclides have been released to the environment, and the ^{41}Ar released to the environment did not exceed 1.85×10^{10} Bq/hr. No accidents occurred during the reactor operations. The environment, including the SPZ, has not been contaminated with RW.

6.2.14 Ukrainian AS Nuclear Research Institute, Kiev

Reactor VVR–M

The core of the VVR–M contains 140 FAs with an initial fuel enrichment of 36%, and the uranium content in each FA is 0.09 kg. The activity of FPs in the core of the operating reactor totals 10^{18} Bq, and the uranium activity is 3.9×10^{8} Bq. The spent fuel discharged from the reactor – 562 FAs with a mass of about 42.3 kg of uranium – is kept in a storage pool. The total activity of FPs is estimated at 10^{17} Bq, and the uranium activity is 1.3×10^{9} Bq.

The reactor is provided with two 330-m^3 metallic reservoirs for storing LRWs. No data are available about the quantity of wastes in these reservoirs. There are two lined concrete bunkers for storing SRWs: one for low-active SRWs and the other for medium-active SRWs. The age of the storage facilities is about 35 years. No data are available on the quantity and activity of the waste contained in the storages facilities.

No radionuclides have been released to the environment, and the ^{41}Ar released to the environment did not exceed 5.55×10^{10} Bq/hr. During the reactor operations, there were no accidents that could have resulted in the RW contamination of the environment, including the SPZ.

6.2.15 Georgian AS Institute of Physics, Tbilisi

Reactor IRT–M

The IRT–M was shut down in 1990 and is being decommissioned. Nearly all of the fuel that has been discharged from the reactor has been transferred to IA Mayak. Some spent FAs of different types (EK–10, IRT–M, TTR) remain in the reactor, where the total content of ^{235}U is equal to 900 g. The uranium burnup is from 3% to 18%. The total activity of FPs is estimated at 10^{14} Bq. There is some fresh (not spent) FAs of EK–10, TTR, IVV–2, and IRT–21 types with a ^{235}U content of 4.23 kg (initial uranium enrichment is 10% and 90%).

No accidents have occurred during operation, and the ^{41}Ar released to the environment did not exceed 3.7×10^{10} Bq/hr. The environment, including the SPZ, has not been contaminated with RWs.

6.2.16 Uzbekistan AS Nuclear Physics Institute, Tashkent

Reactor VVR–SM

The core of the VVR–SM contains 24 FAs with an initial fuel enrichment of 90% and a uranium content of 0.294 kg/FA. The activity of FPs in the core, with the reactor in operation, totals 10^{18} Bq. The uranium activity is 4×10^{8} Bq. The spent fuel that was discharged from the reactor before 1989 was sent to IA Mayak. The spent fuel that was discharged from 1989 until July of 1995 remains in a regular storage pool and in an additional wet storage facility constructed in the 1980s. The age of the storage facility is about 35 years. In both storage facilities (as of July 1995), 143 FAs with a total uranium mass of 24.2 kg were kept. The total activity of FPs is estimated at 10^{17} Bq, and the uranium activity is 1.6×10^{9} Bq.

The reactor has two 300-m^3 steel underground tanks for storing LRW. They contain 200 m^3 of low-active LRWs, with a total activity of 6×10^{10} Bq. The low-active SRWs are stored in metallic casks, kraft paper, or polyethylene bags. Their total mass is 0.5 tons. The medium-active SRWs are kept in an interim repository and their total mass is 0.1 tons. As they build up, the solid wastes are transferred to the regional storage facility in the Republic. The total activity of low- and medium-active RWs is estimated at 7×10^{10} Bq.

No radionuclides have been released to the environment, and the ^{41}Ar released to the environment did not exceed 3.7×10^{10} Bq/hr. No accidents have occurred during the reactor operation. The environment, including the SPZ, has not been contaminated with RW.

6.2.17 Commercial Corporation Foton, Tashkent, Uzbekistan

Reactor IIN

The core of the IIN contains 4.4 kg of uranium enriched to 90%. Due to the pulse mode of operation, the quantity of FPs is insignificant. The total uranium activity is estimated at 3.2×10^{8} Bq.

No radionuclides have been released to the environment. No accidents have occurred. The environment has not been contaminated with RW.

6.2.18 The Alma-Ata Division of the Institute of Atomic Energy of the Kazakhstan National Nuclear Center, Alatan

Reactor VVR–K

The VVR–K was shut down in 1988 in part because of doubts about its seismic stability. After appropriate studies, it was deemed safe for operation. At present, preparations to resume reactor operations are in progress. The spent fuel that was discharged from the reactor from 1974 to 1988 – 332 FAs with a total uranium mass of about 88 kg – is kept in a storage pool. The activity of FPs in the FAs totals about 2×10^{16} Bq, and the uranium activity is 2.7×10^{9} Bq.

The reactor has two 100-m^3 steel tanks for storing low-active LRWs and two 50-m^3 steel tanks for high-active LRWs. The age of the storage facility is 28 years.

No radionuclides have been released to the environment, and the ^{41}Ar released to the environment was about 3.7×10^{10} Bq/hr. No accidents occurred during the reactor operation. The environment, including the SPZ, has not been contaminated.

Critical Test Facility with the Physical Model of VVR–K

The core of the VVR–K physical model contains 59 FAs with 17.9 kg of uranium of 36% enrichment. The total uranium activity is about 6.5×10^{8} Bq.

No radionuclides have been released to the environment. No accidents have occurred at the CTF.

6.2.19 Kurchatov Division of the Institute of Atomic Energy of the Kazakhstan National Nuclear Center, Semipalatinsk District

Reactor IGR

The core of the IGR consists of 2,800 graphite blocks impregnated with uranium of 90% enrichment. Each block contains 3 g of uranium. The fuel burnup, at 2 kWd/block, is insignificant. The total uranium activity is about 6×10^8 Bq.

No radionuclides have been released to the environment. No accidents have occurred in the reactor that could have resulted in RW contamination of the environment, including the SPZ.

Reactor IVG–1

The core of the IVG–1 contains 31 FAs with an initial fuel enrichment of 90%, and each FA contains 0.143 kg of uranium. The fuel burnup, at about 0.1%, is insignificant. The total core uranium activity is 3.2×10^8 Bq.

No radionuclides have been released to the environment. No accidents have occurred in the reactor that could have resulted in RW contamination of the environment, including the SPZ.

6.2.20 Russian Concern Nickel, Norilsk, Russia

Reactor RG–1M

The core of the RG–1M contains 50 FAs with an initial fuel enrichment of 10%, and each FA contains 0.56 kg of uranium. The total activity of FPs in the core of the operating reactor is about 7×10^{15} Bq. The uranium activity is 5.3×10^8 Bq.

No radionuclides have been released to the environment, and the ^{41}Ar released to the environment is insignificant (less than 3.7×10^8 Bq/hr). No accidents have occurred at the reactor.

6.3 Completeness and Availability of Data

The data presented are sufficiently complete. They do not include, however, a number of research reactors, used for special purposes, in particular, reactors in the Federal Nuclear Centers: the All-Russia Research Institute for Experimental Physics (VNIIEF) and the All-Russia Research Institute for Technical Physics (VNIITF). It is important to note that the above-mentioned reactors work in a pulse mode. For this reason, the nuclear fuel in these reactors is not reloaded, and its activity level is low. For this same reason, RW accumulation in the reactors is insignificant.

Data concerning certain research reactors are still classified. For some reactors, the data are not complete because of a lack of necessary measurements. In particular, data concerning isotopic composition of SNF and LRW and SRW are insufficient. No specific databases on RW and RMs accumulated at research reactors in Russia are available.

6.4 Problems of Radioactive Waste and Spent Nuclear Fuel Management at the Research Reactors and in the Research Institutes

All enterprises and institutions that have research reactors in their structure are engaged in the following lines of activities related to RW and SNF management:

- Accumulation, conditioning, and disposal of low- and medium-specific activity-level LRW.
- Accumulation and long-term storage of medium- and high-specific activity-level LRW.
- Accumulation and disposal of SRW.
- Transportation and interim storage of SNF.

The scale of these activities are, naturally, essentially different depending on the research reactors' power, duration of their operation, scope, and subject of experimental research. SNF from research reactors of the FSU was regularly brought to the RT–1 reprocessing plant in the IA Mayak. For the past several years, the economic situation has worsened, and practically no spent fuel has been sent there. Unfortunately, there is no alternative place to send it. Certain problems are connected with the storage of spent experimental fuel as well as the fuel elements that failed in the course of operation. Technical problems regarding the fuel elements' removal can be fully solved, but currently, it is not possible because of a lack of funding.

SRWs are sent for disposal at MNPO Radon, but the situation has worsened there, too. For the past several years, RW removal has been reduced, although no alternative to this solution is foreseen.

The problems that research reactors have with RW and SNF management are the same as those of other enterprises that send their SNF to IA Mayak and their RW to MNPO Radon. Some amount of SNF and radioactive specimens of construction materials from nuclear power plants are sent to research institutes for investigation, increasing the mass of RW in the institutes.

6.5 Conclusion

Research reactors and CTFs built in nuclear research centers differ greatly by design, power level, fuel form and composition, construction materials, and operation mode. On the whole, it should be noted that power level of the reactors is relatively low as compared with commercial NPPs, and comparatively small amounts of RWs have accumulated in the nuclear research centers. Therefore, spent fuel unloaded from research reactors is kept in cooling ponds. The total activity of FPs in spent FAs is about 1.2×10^{18} Bq.

LRW from research reactors are stored in steel and concrete reservoirs, and their total activity is about 10^{15} Bq. The wastes are mainly concentrated in the RF SRC–NIIAR, where the most powerful research reactors of the FSU are operated. In NIIAR, low-level LRWs are pumped into a deep underground repository of a test site after blending, settling, and neutralization. The total activity of LRW injected into the underground repository is 4.1×10^{15} Bq. Low-level SRWs are stored in metallic barrels or paper or polyethylene bags. Medium-level SRWs are kept in interim storage facilities. The total activity of low- and medium-level SRW has been estimated at 5.9×10^{12} Bq. High-level SRWs are transported by special vehicles to special burial facilities with reinforced concrete bunkers. The total activity of high-level SRW is about 3.3×10^{16} Bq. The type of RW is mainly concentrated in NIIAR.

Practically no radionuclides have been released into the environment. During the period of research reactor and CTF operation, there were no accidents that resulted in radioactive contamination of the environment, including the SPZ zone areas.

Bibliography[1]

NIIAR (Research Institute of Atomic Reactor), 1993, *The Research Reactors of NIIAR and Their Experimental Capabilities*, NIIAR, Dimitrovgrad, Russia.

RRC–KI (The Russian Research Center – Kurchatov Institute), 1993, *Research and Development in Reactor Scientific Centers*, RRC–KI, Moscow, Russia.

[1]Much of the information in this chapter is derived from annual reports from research institutes possessing experimental reactors.

Table 6.1. Research reactors of the research centers of the RF and Republics of the FSU.

Reactor type	Fuel characteristics (enrichment, composition, form)	Date of reaching criticality & shutdown	Power, MW		Application
			Initial design	Advanced design	
Russian Research Center – Kurchatov Institute (RRC-KI), Moscow, Russia.					
F–1 (uranium-graphite)	Natural uranium (metal, UO_2, and U_3O_8), uranium of 2% enrichment	1946	0.024	–	Calibration irradiations of activation neutron detectors, tests of new ionization chambers (IC), investigation of IC characteristics, certification of neutron radiation detectors.
VVR–2 (tank-type, water-water)	Uranium of 2% enrichment, UO_2+Al; uranium of 36% enrichment, UAl alloy; 10-mm diameter rods	1954 (1983)	0.3	3.0	Nuclear physics investigations.
RFT (channel-type, graphite)	Uranium of 10–90% enrichment, U+Mg, UO_2+Mg, UO_2+Al, tubes	1952 (1962)	10.0	20.0	Tests of reactor materials, fuel elements (FEs), and fuel assemblies (FAs) for power and research reactors. Nuclear physics investigations.
MR (water-beryllium, channel-type, pool)	Uranium of 90% enrichment, UAl alloy UO_2+Al, tubes)	1963 (1993)	20.0	40.0	Tests of reactor materials, FEs, and FAs for power and research reactors. Neutron radiography. Isotope production.
IRT (pool-type)	Uranium of 10–90% enrichment, UO_2+Al, UAL alloy, tubes	1957 (1979)	2.0	8.0	Nuclear physics and solid-state physics studies. Neutron activation analysis. Isotope production.

Table 6.1. Continued.

Reactor type	Fuel characteristics (enrichment, composition, form)	Date of reaching criticality & shutdown	Power, MW		Application
			Initial design	Advanced design	
IR–8 (pool-type, reconstructed IRT)	Uranium of 90% enrichment, UAl alloy, UO_2+Al, tubes	1981	8	–	Nuclear physics and solid-state physics investigations. Neutron activation analysis. Isotope production. Neutron radiography. Radiation tests of materials.
IIN–3M Gidra (homogeneous, solution)	Uranium of 90% enrichment, water solution of UO_2SO_4	1972	0.01 stationary 2×10^4 (in pulse)	–	Nuclear physics investigations. Neutron activation analysis. FE tests under non-stationary conditions.
GAMMA (water-water, vessel-type)	Uranium alloy, >20% enrichment	1982	0.125	–	FE service life tests.
Argus (homogeneous, solution)	Uranium of 90% enrichment, water solution of UO_2SO_4	1981	0.02	–	Neutron radiography. Neutron activation analysis. Production of isotopes and nuclear filters.
OR (tank-type, water-water)	Uranium of 10% enrichment, UO_2+Al, uranium of 36% enrichment, UAl alloy; 10-mm diameter rods	1989	0.3	–	Investigations and tests of neutron and gamma-radiation shields. Tests of radiation stability of nuclear unit equipment.
Romashka (homogeneous)	Uranium of 90% enrichment, UC_2	1964 (1966)	0.04	–	Investigation of the characteristics of nuclear power units for direct energy conversion.
Topaz–2 (channel-type, uranium Zr-hydride)	Uranium of 90% enrichment, UO_2	1973 (1986)	0.1	–	Investigation of the characteristics of space nuclear power units.

Table 6.1. Continued.

Reactor type	Fuel characteristics (enrichment, composition, form)	Date of reaching criticality & shutdown	Power, MW		Application
			Initial design	Advanced design	
Critical test facility SF–1 (uranium-water)	–	1972	100 W	–	Critical experiments. Investigation of core neutronic parameters of various type water-water reactors.
Critical test facility SF–3 (uranium-water)	Uranium of 90% enrichment, UZr alloy; uranium of 21% enrichment, UO_2	1979	100 W	–	Critical experiments. Investigation of neutronic parameters of water-water reactor cores.
Critical test facility SF–5 (uranium hydride zirconium)	Uranium of 24% and 36% enrichment, intermetallic	1990	100 W	–	Critical experiments. Studies of core physics of the reactor with uranium hydride zirconium fuel.
Critical test facility SF–7 (uranium-water)	Uranium of 80% enrichment, UZr alloy	1975	–	–	Critical experiments. Investigation of neutronic parameters of water-water reactor cores.
Critical test facility Kvant (uranium-water)	Uranium of 90% enrichment, uranium intermetallic	1990	1,000 W	–	Critical experiments. Investigation of neutronic parameters of water-water reactor cores.
Critical test facility with physical model of MR reactor (water-beryllium)	Uranium of 90% enrichment, UAl alloy, tubes	1971	100 W	–	Critical experiments. Simulation of MR core loading. Prereactor investigation of loop experiment parameters.
Critical test facility Delta (uranium-water)	Uranium of 80% and 90% enrichment, UO_2	1985	100 W	–	Critical experiments. Investigation of neutronic parameters of water-water reactor cores.

Table 6.1. Continued.

Reactor type	Fuel characteristics (enrichment, composition, form)	Date of reaching criticality & shutdown	Power, MW		Application
			Initial design	Advanced design	
Critical test facility UG (uranium-graphite, channel-type)	–	1965	–	–	Critical experiments. Studies of physics of uranium-graphite, channel-type reactor cores.
Critical test facility of RBMK (channel-type, uranium-graphite)	Uranium of 2% enrichment, UO_2	1981	25 W	–	Critical experiments. Studies of RBMK core physics.
Critical test facility V–1000 (uranium-water)	Uranium of up to 4.4% enrichment, UO_2	1986	200 W	–	Critical experiments. Studies of neutronic parameters of water-water reactor core.
Critical test facility ASTRA (uranium-graphite)	Uranium of 21% enrichment, UO_2	1981	100 W	–	Critical experiments. Investigation of neutronic parameters of uranium graphite reactor cores.
Critical test facility EFIR–2M (uranium-water)	Uranium of 90% enrichment, UO_2+Al	1973	100 W	–	Critical experiments. Investigation of neutronic parameters of water-water reactor core.
Critical test facility MAYAK (uranium-water)	UAl alloy	1967	10 W	–	Critical experiments. Investigation of neutronic parameters of water-water reactor cores.
Critical test facility NARCISS M2 (uranium hydride zirconium)	Uranium of 96% enrichment, UO_2	1983	10 W	–	Critical experiments. Studies of core physics of the reactor with uranium hydride zirconium fuel.

Table 6.1. Continued.

Reactor type	Fuel characteristics (enrichment, composition, form)	Date of reaching criticality & shutdown	Power, MW		Application
			Initial design	Advanced design	
Critical test facility GROG (uranium-graphite)	Uranium of 7% and 10% enrichment, UO_2.	1980	100 W	–	Critical experiments. Investigations of neutronic parameters of uranium-graphite reactor core.
Critical test facility P (uranium-water)	Uranium of 2.0, 2.4, 3, 3.6, 4.4, 6.5, and 10% enrichment, UO_2	1987	200 W	–	Critical experiments. Investigations of neutronic parameters of water-water reactor cores.
Critical test facility ISKRA (uranium-water)	Uranium of 90% enrichment, UAl alloy, uranium nitride	1996	199 W	–	Critical experiments. Investigations of neutronic parameters of cores of different composition.
Institute of Theoretical and Experimental Physics (ITEF), Minatom, Moscow Russia.					
TVR (heavy water)	Uranium of 80% enrichment, UO_2+Al, tubes	1949 (1986)	0.5	2.5	Nuclear physics and solid-state physics investigations. Neutron physics. Isotope production. Experiments on neutron scattering. Radiation tests.
Critical test facility Mayak (heavy water)	Uranium	1976	1,000 W	–	Critical experiments. Investigation of neutronic parameters of heavy water moderated and cooled reactor cores.
Moscow Institute of Engineering Physics (MIFI), Ministry of General and Professional Education, Moscow, Russia.					
IRT (pool-type)	Uranium of 90% enrichment, UO_2+Al	1967	2	2.5	Investigations of solid-state physics, radiation physics of materials. Radiation physics of semiconductors and dielectrics. Nuclear spectrometry. Reactor physics and engineering.

Table 6.1. Continued.

Reactor type	Fuel characteristics (enrichment, composition, form)	Date of reaching criticality & shutdown	Power, MW		Application
			Initial design	Advanced design	
Research and Design Institute of Power Engineering (NIKIET), Minatom, Moscow, Russia.					
IR–50 (pool-type)	Uranium of 10% enrichment, UO_2+Al; 10-mm diameter fuel rods	1961 (1993)	0.05	–	Experiments on physics of radiation protection. Tests of ion detectors. Neutron activation analysis.
Research Instrumentation Institute (NIIP), Minatom, Lytkarino, Moscow region, Russia.					
IRV–M1 (pool-type)	Uranium of 90% enrichment, UAl alloy, tubes	1974 (1990)	2	–	Radiation tests of electronic instruments.
Konstantinov Institute of Nuclear Physics, Russian Academy of Sciences, Gatchina, Leningrad region, Russia.					
VVR–M (tank-type)	Uranium of 90% enrichment, UO_2+Al, tubes	1959	10	18	Investigations of nuclear and solid-state physics, studies of neutron scattering. Isotope production. Neutron radiography. Tests of fuel rods. Investigations of radiobiology and radiation chemistry.
Critical test facility with physical model of RIK reactor	Uranium of 90% enrichment, rods	1983	–	–	Critical experiments. RIK reactor simulation.
Research Institute of Atomic Reactors (NIIAR), Minatom Ulyanovsk region, Dimitrovgrad, Russia.					
SM–3 (vessel-type)	Uranium of 90% enrichment, UO_2, rods	1961	50	100	Production of transuranium elements. Tests of reactor material samples. Isotope production. Neutron activation analysis.
MIR–M1 (channel-type, pool reactor)	Uranium of 90% enrichment, UO_2+Al, tubes	1966	100	–	Tests of FEs and FAs of power and research reactors. Radioisotope production.

Table 6.1. Continued.

Reactor type	Fuel characteristics (enrichment, composition, form)	Date of reaching criticality & shutdown	Power, MW		Application
			Initial design	Advanced design	
BOR–60 (fast reactor)	Uranium of 90% enrichment, UO_2, plutonium, rods	1969	100 W	–	Tests of reactor materials and fuel rods with different fuel types.
RBT–6 (pool-type)	Uranium of 63% enrichment, UO_2, rods (spent FA of SM–3)	1975	60	–	Tests of reactor material samples. Neutron radiography.
RBT–10/1 (pool-type)	Uranium of 63% enrichment, UO_2, rods (spent FA of SM–3)	1983	6	–	Tests of reactor material samples.
RBT–10/2 (pool-type)	Uranium of 63% enrichment, UO_2, rods (spent FA of SM–3)	1984	10	–	Tests of reactor material samples.
VK–50 (vessel-type, boiling)	Uranium enriched to 2, 2.4, and 3%, UO_2, rods	1963 (1989)	10	–	Generation of electricity and heat. Investigation of the technology of boiling, vessel-type reactors.
ARBUS (vessel-type with an organic coolant)	Uranium of 90% enrichment	1963 (1989)	12	–	Investigation of organic coolants of various types.
Critical test facility with the physical model of SM–3 reactor	Uranium of 90% enrichment, UO_2, rods	1961	100 W	–	Critical experiments. Simulation of SM–3 core loading. Prereactor investigation of the parameters of ampoule and loop facilities.

Table 6.1. Continued.

Reactor type	Fuel characteristics (enrichment, composition, form)	Date of reaching criticality & shutdown	Power, MW Initial design	Advanced design	Application
Critical test facility with the physical model of MIR reactor	Uranium of 90% enrichment, UO_2+Al, tubes	1966	100 W	–	Critical experiments. Simulation of MIR–M1 core loading configurations. Prereactor investigations of the parameters of loop experiments.
Nuclear Physics Research Institute (NIIYaF), Ministry of General and Professional Education, Tomsk, Russia.					
IRT–T (pool-type IRT)	Uranium of 90% enrichment, UO_2+Al, tubes	1967	2	6	Neutron activation analysis. Neutron doping of silicon.
Branch of the Research and Design Institute of Power Engineering (NIKIET), Minatom, Zarechny, Sverdlov region, Russia.					
IVV–2M (pool-type)	Uranium of 90% enrichment, UO_2+Al, tubes	1966	10	15	Tests of FR and reactor materials. Installations of direct energy conversion. Solid state physics investigations. Neutron scattering studies.
Institute of Physics and Power Engineering (FEI), Minatom, Obninsk, Russia.					
BR–10 (fast, sodium)	Uranium of 90% enrichment, rods	1958	5	8	Tests of fuel rods and reactor materials. Nuclear physics and radiobiology investigations. Neutron activation analysis. Radioisotope production.
AM (channel-type, graphite)	Uranium of 4.4% and 10% enrichment	1954	30	10	Tests of fuel rods and reactor materials. Installations of direct energy conversion. Radioisotope production.
Critical test facility BR–1	–	1955	100 W	–	–

Table 6.1. Continued.

Reactor type	Fuel characteristics (enrichment, composition, form)	Date of reaching criticality & shutdown	Power, MW		Application
			Initial design	Advanced design	
Critical test facility BFS–1	U, Pu, UO_2	1961	200 W	–	Critical experiments. Investigation of neutronic parameters of fast reactor cores.
Critical test facility RF–GS	Uranium of 90% enrichment, uranyl nitrate solution	1962	–	–	Critical experiments. Investigation of neutronic parameters of various-type reactor cores.
Critical test facility MATR–2 (uranium-water)	–	1963	–	–	Critical experiments.
Critical test facility PS–2	–	1969	200 W	–	Critical experiments.
Critical test facility T–2	Uranium of 90% enrichment	1965	2 W	–	Critical experiments.
Critical test facility SGO	Uranium of 90% enrichment, UO_2, U–Be, rods	1968	100 W	–	Critical experiments.
Critical test facility Strela	Uranium of 90% enrichment, UN, balls, blocks, bushes	1968	100 W	–	Critical experiments.
Critical test facility BFS–2	U, Pu, UO_2	1969	100 W	–	Critical experiments. Investigation of the neutronic parameters of fast reactor cores.
Critical test facility KOBR	–	1970	–	–	Critical experiments.

Table 6.1. Continued.

Reactor type	Fuel characteristics (enrichment, composition, form)	Date of reaching criticality & shutdown	Power, MW		Application
			Initial design	Advanced design	
Critical test facility AMBF–2	Uranium of 2% enrichment, magnesium intermetallic	1984	–	–	Critical experiments.
Critical test facility FS–1M	–	1985	2 W	–	Critical experiments.
Critical test facility K–1	–	1989	–	–	Critical experiments.
Karpov Institute of Physical Chemistry, Ministry of Industry, Obninsk, Kaluga region, Russia.					
VVR–C (tank-type VVR)	Uranium of 30% enrichment, UAl alloy, UO_2+Al, tubes	1964	15	–	Investigations in radiation chemistry. Radiation tests of electronics. Neutron radiography, radioisotope production. Neutron doping of silicon.
Joint Institute of Nuclear Research, Dubna, Moscow region, Russia.					
IBR–30 (pulse)	Uranium of 90% enrichment, Pu, rods	1969	0.03 (avg.) 90 (in pulse)	–	Investigations of nuclear, solid-state, and neutron physics.
IBR–1 (pulse)	PuO_2, 8.6-mm diameter rods	1980	4 (avg.) 8.3×10^3 (in pulse)	–	Investigations of nuclear, solid-state, and neutron physics.
LAS Institute of Physics, Riga district, Latvia.					
IRT–M (pool-type)	Uranium of 90% enrichment, UAl, UO_2+Al, tubes	1961	2	5	Investigations of solid-state physics, nuclear spectroscopy, and radiobiology. Neutron radiography. Radiation tests. Neutron activation analysis.

Table 6.1. Continued.

Reactor type	Fuel characteristics (enrichment, composition, form)	Date of reaching criticality & shutdown	Power, MW		Application
			Initial design	Advanced design	
BAS Nuclear Power Institute, Sosny, Minsk district, Belorussia.					
IRT–M (pool-type)	Uranium of 90% enrichment, UAl alloy, tubes	1962 (1988)	2	4	Studies of basic principles of the use of a dissociating coolant. Investigations of solid-state physics, radiation chemistry, and radiobiology. Neutron activation analysis.
UAS Nuclear Research Institute, Kiev, Ukraine.					
VVR–M (tank-type)	Uranium of 36% enrichment, UAl alloy, UO_2+Al, tubes	1960	10	–	Investigations of nuclear and solid-state physics, neutron scattering, and radiobiology. Radiation tests of metals, alloys, and semiconductors. Radioisotope production. Neutron radiography.
GAS Institute of Physics, Tbilisi, Georgia.					
IRT (pool-type)	Uranium of 90% enrichment, UAl alloy, tubes	1959 (1990)	2	8	Investigations of radiation physics of solids, radiation chemistry, and neutron scattering. Neutron activation analysis of biologial objects.
URAS Nuclear Physics Institute, Tashkent, Uzbekistan.					
VVR–SM (tank-type VVR)	Uranium of 90% enrichment, UAl alloy, UO_2+Al, tubes	1959	2	10	Investigations of nuclear and solid-state physics. Radiation test of materials and facilities. Radioisotope production. Neutron activation analysis.
Commercial Corporation Foton, Tashkent, Uzbekistan.					
IIN (homogeneous, solution)	Uranium of 90% enrichment, aqueous solution UO_1SO_4	1975	1.5 (in pulse)	–	Neutron activation analysis.

Table 6.1. Continued.

Reactor type	Fuel characteristics (enrichment, composition, form)	Date of reaching criticality & shutdown	Power, MW		Application
			Initial design	Advanced design	
Alma-Ata Division of the Institute of Atomic Energy, Alma-Ata district, Kazakhstan.					
VVR–K (tank-type)	Uranium of 36% enrichment, UAl alloy, tubes	1967 (shut-down in 1988 for strengthening of seismic stability)	10	–	Radiation tests of materials and nuclear power unit components. Investigations of neutron scattering.
Critical test facility with a physical model of VVR–K	Uranium of 36% enrichment, UAl alloy, tubes	1972	100 W	–	Modeling of core loading in VVR–K with loops and ampoule of test conditions in loop channels.
Kurchatov Division of Institute of Atomic Energy, Semipalatinsk district, Kazakhstan.					
IGR (uranium-graphite)	Uranium of 90% enrichment	1961	5.1 MJ (total energy yield)	–	Investigation of FE and FAs behavior under unsteady conditions.
IVG–1	Uranium of 90% enrichment	1972	60	–	Tests of various type fuels.
Norilsk Mining and Metallurgy Combine, Russian Concern Nickel, Norilsk, Russia.					
RG–1M (tank-type, water-water)	Uranium of 10% enrichment, UO_2+Al, 10-mm diameter rods	1970	0.1	–	Neutron activation analysis.

Chapter 7

Plutonium Production and Radiochemical Reprocessing of Spent Nuclear Fuel

7.1 Introduction

In the former Soviet Union (FSU), the production of weapons-grade plutonium was concentrated at three enterprises: the Industrial Association (IA) Mayak in Ozersk, Chelyabinsk region; the Siberian Chemical Combine (SCC) in Seversk, Tomsk region; and the Mining and Chemical Combine (MCC) in Zheleznogorsk, Krasnoyarsk region. All the enterprises are located on the territory of Russian Federation (RF). The process of extracting plutonium from irradiated uranium involves separating the two elements and purifying them of fission products. Metallic plutonium articles are the final product of the process. The nuclear materials production is accompanied by the generation of radioactive wastes (RWs), which undergo suitable processing and then are sent to storage.

7.2 Characteristics of Accumulated Radioactive Wastes

Irradiated materials reprocessing, like any chemical production work, generates waste that must be disposed. RWs are differentiated by their aggregate state, specific (volumetric) activity level, and origin. Additionally, the

RWs are divided into liquid, solid, and gaseous wastes according to their aggregate state. To the category of liquid radioactive wastes (LRWs), in addition to waste process solutions, various suspensions and sludges are added. Solid radioactive wastes (SRWs) include metals, concrete, wood, organic films, work clothes, etc. Gaseous wastes may be provisionally subdivided into two groups: gases containing induced (in nuclear reactors) activity and gases evolving from irradiated uranium reprocessing and further chemical and metallurgical treatment of radioactive materials (RMs). Currently, high-, medium-, and low-level RWs are kept at the enterprises reprocessing irradiated uranium. According to their origin, RWs may be subdivided into process wastes (reactor wastes and the wastes from radiochemical, chemical, metallurgical, UF_6 production, uranium enrichment, and other production work) and nonprocess wastes (wastes of sanitary treatment points, special laundries, research laboratories, and other auxiliary work). In terms of total activity, the process wastes dominate. Most (practically, more than 99%) of the fission products are concentrated in LRWs, since aqueous processes were and are now used for the industrial reprocessing of irradiated uranium.

7.2.1 The IA Mayak

The IA Mayak is situated on 160 km^2, with the main production site located in the central and eastern parts of the territory. Dwelling blocks are arranged in the northwestern and southwestern parts of the territory. The area between the production and living zones is occupied by auxiliary production facilities. The nearest settlements to the industrial zone are 12 km to the west (the town of Kyshtym), 16 km to the north (the town of Kasli), and 5 km to the south (the Novogorny settlement). There are also some small villages in the area around the IA Mayak territory, 10 km away. The sanitary-protection zone (SPZ) to the northeast and west is 10 km long. It widens to 18 km on the eastern (leeward) side, where it borders habitated areas. The total area of the SPZ is 250 km^2, and includes bodies of water with a surface area of 60 km^2 and 150 km^2 of forests. In the SPZ, the enterprise is allowed to conduct restricted agricultural and forestry activities.

A sanitary observation zone has been established around the industrial zone of the enterprise (a 40-km radius), and within the 1957 radioactive trace area along the border of the Sverdlovsk region. Its total area is 1,800 km^2, including bodies of water with a surface area of 216 km^2

and 900 km^2 of forests. Within the observation zone territory are 38 settlements inhabited by 150,000 people. The general scheme of the Mayak territory is shown in *Figure 7.1*.

Two natural water reservoirs on the IA Mayak territory are used for industrial circulating water supply, and two artificial water basins that were originally swamps are used for storing medium-level LRWs.

The production complex of IA Mayak includes:

- Production reactors.
- Radiochemical plant: for the reprocessing of irradiated materials in order to separate and purify uranium and plutonium salts.
- Radioisotopes production.
- Chemical and metallurgical plant: for fabrication of metallic uranium and plutonium.
- Chemical production works.

Additionally, a number of auxiliary facilities are operated at the IA Mayak:

- Water treatment plant.
- Repair and mechanical plant.
- Storehouses of combustible and lubricating materials.
- Motor car transportation and railway facilities.
- Construction divisions.
- Objects for RW processing, storage, and disposal.

The first uranium-graphite reactor, the 100-MW "A," was started up on 19 June 1948, and it was decommissioned on 16 June 1987. Later, a number of uranium-graphite reactors were put into operation at the IA Mayak, including the AV-1 (15 June 1950), AV-2 (6 April 1951), AI (22 December 1951), and AV-3 (15 September 1952).

The first heavy-water reactor, the 100-MW OK-180, was put into operation on 17 October 1951. The OK-180 reactor was decommissioned in 1965 and then disassembled.

The next heavy-water reactor, OK-190, was put into operation on 27 December 1955. The reactor was operated for 10 years and was shut down on 8 November 1965. Later, the reactor was disassembled, and in its place another reactor (OK-190m) was started up in April 1966. The OK-190m

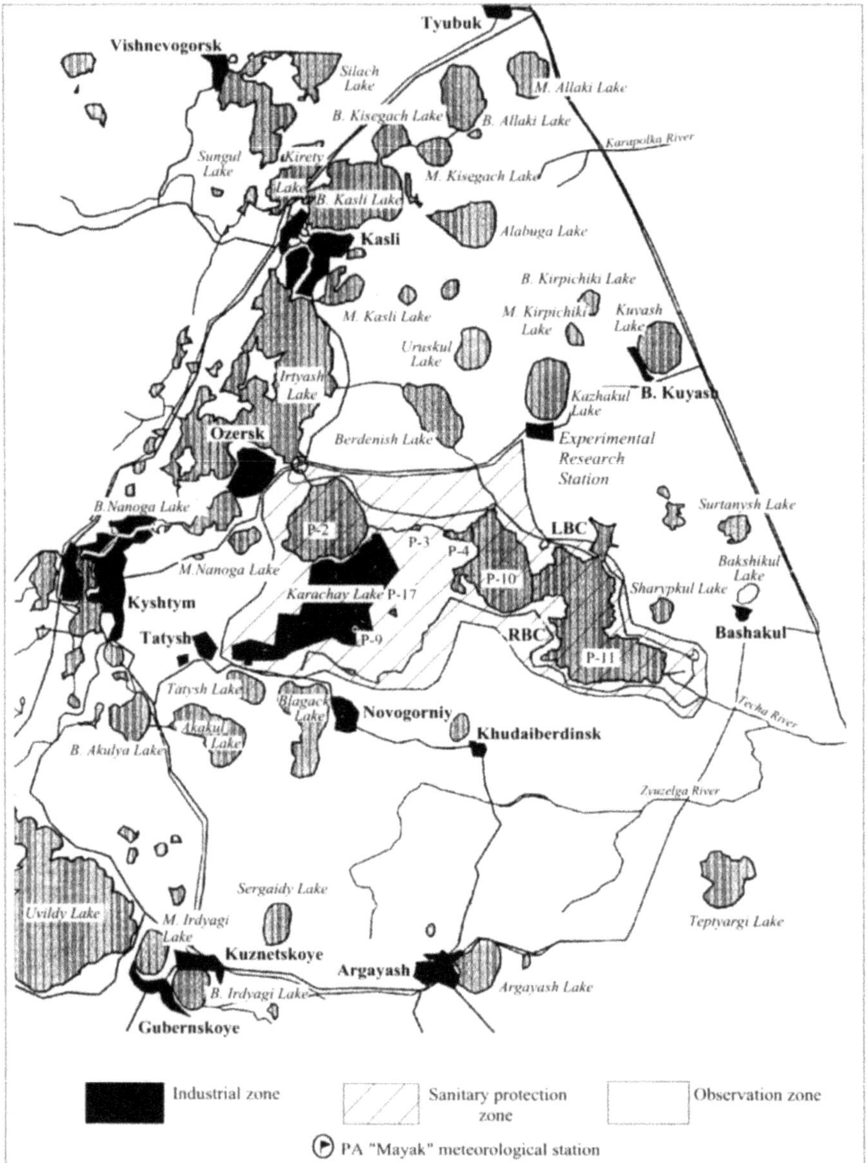

Figure 7.1. General scheme of the IA Mayak territory.

reactor operated for 20 years, and was shut down in 1986. Afterward, reactors that ensured more reliable operation were built: The Ruslan reactor was started up on 16 June 1979, and then the Ludmila reactor was put into operation in 1980.

To separate plutonium generated in nuclear reactors from uranium and fission products, a radiochemical plant was built at the IA Mayak and put into operation in late 1948. Irradiated uranium slugs covered with aluminum cladding entered the plant and, after removal of the cladding, were dissolved in nitric acid. The solution – containing uranium, plutonium, and fission products – was processed using the well-known acetate-fluoride technology. The experience gained from the use of this technology allowed three new radiochemical plants to begin operation in 1952. At these plants, irradiated uranium slugs were processed with the improved acetate scheme. Later at the plants, sorption decontamination of plutonium concentrates was implemented using locally made ion-exchange resins.

In spite of improvements in the technological process, the acetate technology had a very serious disadvantage: Large volumes of LRWs were generated. As a result of many years of investigations conducted at research institutes and at the IA Mayak's central laboratory, a compact extraction–sorption technological process was designed and suggested for use. Radiochemical production started using the new technology based on the process.

Reprocessing of spent nuclear fuel (SNF) has been carried out at the RT-1 radiochemical plant since 1971. The reprocessed SNF includes fuel from nuclear power plants (NPPs) with VVER-440 and BN-600 reactors, naval power reactors, and research reactors. The plant's production capacity is 400 tons per year. Twenty years of experience at the RT-1 plant has proven the technological process and equipment to be highly reliable. The closed nuclear fuel cycle concept was first realized in the USSR at Mayak's uranium line: After purification, uranium in the form of fused uranyl-nitrate is supplied to plants for RBMK reactors' fuel manufacturing. Plutonium separated as a result of the regeneration process in the form of dioxide is sent for storage. It is completely suitable for making mixed uranium-plutonium fuel for fast reactors.

Construction of new shops for the chemical and metallurgical plant was started in 1948, and conducted in two stages. The first stage was designated for processing radiochemical plant product solutions into metallic plutonium objects used in nuclear warheads. The second stage, which also was designated for metallic uranium fabrication, was put into operation after the first lots of highly enriched uranium had been produced.

RW Management at the IA Mayak

The first production reactor, radiochemical plant, and other production facilities of IA Mayak were situated in the area between the Techa and Mishelyak rivers. Slightly active reactor waters were discharged into Kyzyl-Tash Lake (water reservoir P-2). Until 1953, water flowed into and out of the lake. From 1953 to 1956, flood waters from the Irtyash-Kasli system periodically passed through the lake. Water reservoir P-2 has been closed since 1957, and it was used as a heat sink in the system of circulating water supply of nuclear reactors. In summer, to keep the necessary water level, water reservoir P-2 is replenished from Irtyash Lake.

In the early period of the operation of IA Mayak, radionuclides from both the radiochemical plant and single-pass uranium-graphite production reactors (A, AI, AV-1, 2, 3) were released into the Techa River, running from Kyzyl-Tash Lake. However, the discharge of about 1.11×10^{17} Bq from March 1950 to late 1951 was a contribution of only the radiochemical plant. Some of the plant's radioactive discharges were anticipated and allowed for by the project, but mainly they resulted from unforeseen emergency situations.

In 1949, the high-level waste storage facilities at the radiochemical plant were filled to capacity, jeopardizing further operation. At that time, it was decided to discharge the wastes into the Techa River. Since 28 October 1951, the main part of the discharges has been directed to Karachay Lake, and only $(3.7\text{–}7.4) \times 10^{12}$ Bq/day have been disposed into the Techa River. The discharge of medium-level LRWs from the radiochemical plant from 1949 to 1951 has led to radioactive contamination of bottom sediments and the river's upper flood plain. To prevent the radionuclides from spreading downstream, a cascade of water reservoirs (P-3, P-4, P-10, and P-11) was created.

Water reservoir No. 3's dam was built in 1951, 2.5 km from water reservoir P-2. Regeneration solutions from clean-up facilities used to reprocess wastewaters were disposed into the water reservoir. An operating spillway to water reservoir P-4 is carried out over dam No. 3.

The dam of water reservoir P-4 is situated 3 km downstream from dam No. 3. The dam existed earlier, and in 1956 it was built up and reinforced.

Regeneration solutions of the water treatment plant are disposed to water reservoir P-4. An operating spillway to water reservoir P-10 is carried out over dam No. 4.

The dam of water reservoir P-10 was built in 1956, 5 km downstream from dam No. 4 on the Techa River. It has an operating spillway to water reservoir P-11.

Dam No. 11 was built in 1964, 17 km downstream from dam No. 10 on the Techa River. Filling of water reservoir P-11 was started in 1966. The water body locks up the cascade of water reservoirs, and it has no spillway. Below dam No. 11, a drainage system was made to collect water that filters through the dam's body. The seepage waters can be returned to water reservoir P-11, if permissible levels of radionuclides concentration are exceeded.

The entire cascade of water reservoirs is designated for low-level LRW storage. The waters are discharged directly into water reservoirs P-3 and P-4, from which, due to natural water overflow, the radionuclides and salts move into water reservoir P-10 and then into water reservoir P-11. To collect on-surface run-off from the water catchment area, interceptor canals were built along the water reservoirs of the cascade – the left bank (north) canal and the right bank (south) canal – which discharge into the Techa River below dam No. 11. The water reservoirs of the cascade are filled from the IA Mayak discharges, atmospheric precipitation, on-surface run-off from the limited water catchment area, and the influx of groundwater.

Water reservoir P-6 is Tatysh Lake and has no run-off. At present, it is used for circulating water supply. Besides the cooling waters, nonprocessed and household wastewaters from the industrial area come into the water body after cleaning.

Water reservoir P-17 is an engineered structure, made in a natural cavity as a result of the construction of earthen dam No. 17 in 1949. The water body is used for storage of medium-level LRWs. At present, a project for its complete elimination is being developed.

Water reservoir P-9 is Karachay Lake. Since 1951 it has been used for medium-level LWR storage. Since 1988, work aimed at eliminating the water reservoir has been conducted.

The Characterization of Accumulated RMs and RWs

Reprocessing 1 ton of irradiated uranium slugs during the initial period of the radiochemical plant operation used about 25 tons of acidic and alkaline solutions and more than 2,000 tons of cooling and process water. The amount of salts generated in the process was approximately 15 tons.

In the RT-1 plant, the reprocessing of 1 ton of SNF from the VVER-440 reactors resulted in the generation of 45 m^3 of high-level, 150 m^3 of medium-level, and 2,000 m^3 of low-level LRWs.

After 1996, all the plant's high-level wastes, after evaporation, were solidified by vitrification. Medium-level wastes will be processed by bitumization, and the low-level wastes, after ion exchange cleaning to the maximum permissible concentrations, will be discharged into open-water reservoirs. The data on amounts of accumulated LRWs and SRWs, as well as characteristics of their storage sites, are presented in *Tables 7.1* and *7.2*.

The Consequences of IA Mayak Activities

The IA Mayak activities led to intensive radioactive contamination of the Urals area – with the highest levels in the northern part of Chelyabinsk region. It was caused by imperfection of nuclear technologies at the initial stages of the enterprise's operation and a lack of knowledge concerning the behavior of radioactive substances in the environment. The most significant radioactive contamination of the environment and exposure of the population to radiation took place during the first half of the time period of IA Mayak's operation.

Due to a lack of proper systems for the management of the radiochemical production work's LRWs and an oversimplified approach to the problem of disposal of such wastes into the natural aquatic systems from 1949 to 1956, wastes with a total activity of 10^{17} Bq were discharged into the small Techa River, which flowed into the Iset River. As a result, 124,000 people living along the river in the Chelyabinsk and Kurgan regions were exposed to radiation.

In autumn 1957, design defects of the first high-level LRW storage reservoirs led to radiation overheating of a tank followed by an explosion of nitrate-acetate salts contained in it. The explosion caused about 7.4×10^{17} Bq of radioactive products to be ejected. Of this amount, 90% fell in the vicinity of the tank, whereas the remaining activity (7.4×10^{16} Bq) formed a kilometer-high radioactive cloud. This radioactivity was scattered a significant distance by wind and led to radioactive contamination of the northern part of the Chelyabinsk region and the southern part of the Sverdlovsk region. The contaminated zone, later

Table 7.1. Amount of accumulated SRWs and LRWs, and characteristics of their storage sites at the IA Mayak.

Description of storage site	Period of operation		Volume, 1,000 m^3		Area, $10^3 m^2$	RW activity, Bq	Notes
	Start	End	Design	Actual			
SRWs							
Concrete structures divided into compartments; concrete compartments lined with steel. Reinforced concrete waterproofed storehouses. Vitrified materials' cooled storehouse of honeycomb type. 18 objects.	1949		270	53	19.55	1.06×10^{19}	SRWs, contaminated with ^{90}Sr, ^{137}Cs, ^{134}Cs, ^{235}U, ^{238}U, ^{238}Pu, ^{241}Pu
Concrete and earthen trenches and natural cavities to store building garbage, wiping materials, plastic film, work clothes, etc.; filled structures are covered with soil. Platforms paved with asphalt. 93 objects.	1949		–	398	–	1.11×10^{16}	Metallic, concrete, wood, organic films wastes, etc., contaminated with ^{90}Sr, ^{137}Cs, ^{134}Cs, ^{235}U, ^{239}Pu
SRWs in total			–	451	–	1.06×10^{19}	
LRWs							
Stainless steel tanks to store suspensions. Stainless steel cooled tanks for intermediate storage of solutions received from the spent nuclear fuel reprocessing plant. 5 objects.	1950		–	30.7	–	1.42×10^{19}	High- and medium-specific activity level sludges and solutions containing ^{90}Sr, ^{137}Cs, ^{134}Cs, ^{103}Ru, ^{106}Ru
Medium-level LRW discharges Water reservoir No. 9 – natural with filtering bottom. Water reservoir No. 17 – natural with filtering bottom.	1949 1951		–	220	320	4.44×10^{18}	Water, containing ^{90}Sr, ^{137}Cs, ^{239}Pu, ^{235}U, ^{3}H
Low-level LRW discharges after cleaning facilities Water reservoirs Nos. 3, 4, 10, 11 – engineered with filtering bottom.	1950		–	850	820	6.29×10^{14} Total activity of sediments 3.7×10^{15} Bq	Water, containing ^{90}Sr, ^{137}Cs, ^{144}Ce, ^{3}H
Water reservoirs Nos. 2 and 6 – natural with filtering bottom, designated for circulating water supply systems of production reactors' cooling.	1948		–	81,400	18,600	8.14×10^{14}	Water, containing ^{90}Sr, ^{106}Ru, ^{239}Pu, ^{241}Am, ^{3}H
LRWs in total			–	82,500	19,740	1.87×10^{19}	
RWs in total			–	–	–	2.93×10^{19}	

Table 7.2. Contamination of water reservoirs at the IA Mayak.

Water reservoir	Area, km^2	Volume, m^3	In water, Bq/l		In sediments, Bq/l	
			Sr-90	Cs-137	Sr-90	Cs-137
P-2	19.00	83.00	4.1×10^{2}	1.7×10^{2}	4.8×10^{4}	1.9×10^{6}
P-6	3.60	17.50	13.7	7.4×10^{-1}	1.1×10^{4}	1.4×10^{6}
P-9	0.25	0.40	6.3×10^{7}	4.4×10^{8}	1.1×10^{10}	5.2×10^{10}
P-10	16.60	76.00	1.3×10^{4}	3.2×10^{2}	1.3×10^{5}	5.6×10^{9}
P-11	41.00	217.00	1.9×10^{3}	7.4×10^{-1}	4.8×10^{4}	4.8×10^{3}
P-17	0.17	0.33	2.6×10^{5}	1.5×10^{5}	0.4×10^{10}	1.2×10^{9}

called East-Urals radioactive trace (EURT), includes about 20,000 km^2 within the limits of the minimum detectable radioactive contamination level of ^{90}Sr (0.1 Ci/km^2), and 1,000 km^2 in the limits of a contamination density of 2 Ci/km^2 of ^{90}Sr (*Figure 7.2*). The latter figure was adopted as the permissible level of safe exposure. At the time, 272,000 people lived in the contaminated territory. Maps of the territories contaminated with ^{90}Sr and ^{137}Cs were drawn up in accordance with data from investigations conducted in 1993 by the Chelyabinsk Regional Center on Hydrometeorology and Environmental Monitoring (see *Color Plates 2* and *3*).

After cessation of discharges into the Techa River, the medium-level LRWs of 1953 were disposed into the small (0.45 km^2), shallow, and swampy Karachay Lake. In spring 1967, the lake's banks were exposed due to extreme drought conditions. Over a two-week period, about 2.22×10^{13} Bq of radioactive products were scattered by gusty winds over the adjacent territory. An area of 1,800 km^2 was within the limits of contamination density of 0.1 Ci/km^2 of ^{90}Sr. The contamination was essentially lower level, compared to the 1957 explosion, and fell mainly on the EURT territory in the limits of Chelyabinsk region. Approximately 40,000 people underwent additional radiation exposure.

Another source of environmental contamination resulting in population exposure connected with IA Mayak activities were the regular releases of process radioactive products into the atmosphere from the plants' smokestacks. The greatest releases, caused by the growth of the production capacities and imperfection of the gas clean-up systems, took place during the first 10 years of IA Mayak's operation.

As of 1 January 1996, 2,736 km^2 of land have been contaminated with ^{137}Cs and ^{90}Sr (see *Table 7.3*).

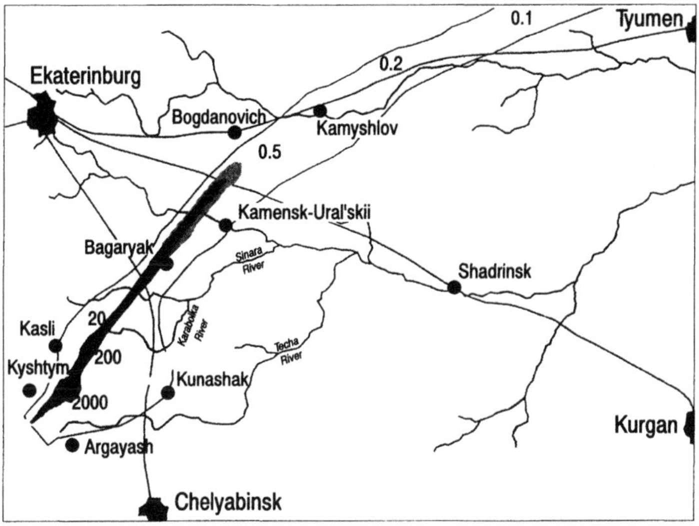

Figure 7.2. The area of the East-Urals radioactive trace. The figures on the scheme correspond to the initial density of ^{90}Sr contamination, Ci/km^2.

Table 7.3. Contaminated lands at the IA Mayak.

Distribution of the contaminated land areas by exposure rate level, μR/hr	Contaminated lands area, km^2			
	Including the territories of			
	Production site	SPZ	Observation zone	Total
Up to 60	–	–	1,055	1,055
60–120	–	–	888	888
120–240	21	31	–	52
240–1,000	15	97	390	502
>1,000	21	71	147	239
Total	57	199	2,480	2,736

7.2.2 The SCC

The construction of the SCC was started in March 1949. The production site is on the second large flood-plain terrace of the right bank (east) of the Tom River, 12–15 km to the north of the city of Tomsk. The local relief is sloped on the west side toward the river. There are many small rivers and streams flowing to the west on the terrain. The river side part of the area is swampy and falls to the main river behind the embankment. The left bank (west) of the river is low.

The SCC includes the following production works:

- Reactors: for plutonium production and electric and heat power generation.
- Radiochemical plant: for the reprocessing of irradiated materials in order to separate and purify uranium and plutonium salts.
- Chemical and metallurgical plant: for the fabrication of metallic uranium and plutonium.
- Sublimate plant: for U_3O_8 and UF_6 production.
- Isotope enrichment plant: for enriched uranium production.
- Nuclear fissile materials storehouses and facilities: for storage of uranium oxides, UF_6, articles of metallic uranium having different enrichment, irradiated production reactors' standard uranium slugs, plutonium oxides, and articles of metallic plutonium.
- Objects for RW processing, storage, and disposal.
- Thermal electric plant: for electric and heat power generation.

The construction of the uranium enrichment plant was started in 1951. The first stage of the plant was put into operation on 26 July 1953. Until 1973, gas diffusion technology was used for uranium isotope separation. Later, the plant converted to highly efficient centrifuge technology. At present, because of a reduction in the state's demand for enriched uranium, the separation capacities of the plant are not fully utilized. This has allowed the plant to render commercial services to foreign firms interested in uranium enrichment.

The construction of the first production uranium-graphite reactor, the I-1, was started in 1952. It was put into operation on 20 November 1955, and went through several stages of modernization. It operated for 35 years and was shut down on 21 August 1990 because of the reduction of weapons-grade plutonium production.

In 1954, a new type of power and production complex known as the Siberian Nuclear Plant was designed and installed. In September 1958, the first stage of the NPP-1, based on the 100-MW, EI-2 uranium-graphite reactor, was put into operation. The third reactor, AD-3, was started up on 14 July 1961. The second and the third reactors were stopped on 31 December 1990 and on 14 August 1992, respectively. In 1959, construction of two uranium-graphite reactors, ADE-4 and ADE-5, was started. The reactors began operating in 1965 and 1967, respectively, and they are still operating. The I-1 reactor, which had a single-pass system for its core cooling, discharged cooling water into the Tom River. The other four reactors were constructed with closed cooling systems.

The UF_6 production (sublimate) plant was constructed in 1951. In April 1954, the anhydrous hydrogen fluoride production department was put into operation, and within a year the UF_4 production department was started up. The plant has been modernized, which has allowed its production capacity to increase by several times while substantially decreasing its releases and disposal of radionuclides and toxic chemical substances.

The radiochemical plant was constructed from 1953 to 1961. The first stage of the plant began operation in August 1961, and the second stage began operation in October 1962. The reprocessing of irradiated uranium slugs was conducted using the acetate precipitation scheme. In 1983, the plant converted to a high-capacity extraction technology allowing the analogous production works at the IA Mayak to stop, and the volumes of RWs decreased significantly.

In 1958, the decision was taken to build a chemical and metallurgical plant at the SCC, and in August 1961 the plant produced its first products.

RW Management at the SCC

As a result of reprocessing of irradiated uranium at the SCC, great amounts of high-, medium-, and low-level RWs have accumulated. Solutions and sludges are included in LRWs. All the wastes are divided into processed and nonprocessed. The main amount of liquid processed wastes (more than 95%) are generated at the radiochemical plant. Disposal of the wastes into deep underground strata is the main method of LRW management at the SCC. The high-level wastes undergo intermediate storage in stainless steel reservoirs, and then, after corresponding treatment, they are sent to underground disposal. The medium-level wastes, after corresponding treatment, are also sent to underground disposal. For intermediate storage of LRWs, both open on-surface storage facilities and special closed-type storage facilities are used. Medium-level process wastes of the chemical and metallurgical production works are stored in an open pool. Processed wastes of the enrichment plant are added to the wastes sent for underground disposal. LRWs of the sublimate plant are sent to two sludge-storage facilities.

Other LRWs and sludges are classified according to the method of their processing. The low-level wastes of all plants are sent to two water reservoirs, from which they are sent to clean-up facilities for processing. Up to 50% of the wastewaters from the clean-up facilities undergo underground disposal. Another part, after being cleaned to standard levels (coagulation, mechanical cleaning, ion exchange), is discharged into the river through an intermediate water reservoir-settler.

The LRWs that are not subjected to processing are sent to open sludge-storage facilities. This category mainly includes sludges, filtering materials, and regenerates coming from clean-up facilities.

The methods of SRW management depend on the waste type. Depending on the specific activity level, they are either disposed into earthen or concrete burials, or piled up in special compartments at sites of regular storage. SRWs from enrichment and uranium production plants are stored and disposed at the plants' production sites. Since 1961, a specially equipped facility has been used for burial of SRWs of the radiochemical, chemical, and metallurgical plants. Low-level SRWs are disposed in trench-type burials, and medium- and high-level wastes are stored in concrete structures.

LRW Storage Facilities

Open storage facilities and underground disposal sites are used to manage LRWs. To process LRWs at the SCC, an open storage facility, two underground disposal sites, a low-level waste-processing surface site, and a station for medium- and high-level waste treatment are used.

Basins B-1, B-2, and B-25

At the first stage of the radiochemical plant operation, medium-level LRWs were discharged into the B-1 basin and then into the B-2 basin. Afterward, as the underground LRW disposal site was put into operation, the basins were used as reservoirs for intermediate storage of solutions before pumping them into the underground strata. They have since been taken out of service. The total activity of long-lived nuclides collected there is estimated at 4.66×10^{18} Bq. The reservoirs are not large in volume, and during summer droughts evaporation of some of the water is possible.

The B-1 basin is an on-surface storage facility with a 1-m-thick loam wall at the bottom and along the slopes. The wall is covered with a 1-m-thick soil layer. Its area is 60,300 m^2, its capacity is 150,000 m^3, and its actual filled volume is 110,000 m^3. The total activity of radionuclides, collected in the basin, is estimated at 2.7×10^{18} Bq. At present, work on decommissioning the basin is being conducted.

The B-2 basin is an on-surface storage facility with a 1-m-thick loam wall at the bottom and along the slopes. The wall is covered with a 1-m-thick soil layer. Its area is 51,400 m^2, its capacity is 135,000 m^3, and its actual filled volume is 63,700 m^3. The collected radionuclides activity totals 1.87×10^{18} Bq. At present, work on decommissioning the basin is being conducted. Work was started in 1991 to fill the B-2 basin with soil.

The B-25 basin is an on-surface storage facility with a 1-m-thick loam wall at the bottom and along the slopes. The wall is covered with a 1-m-thick soil layer. Its area is 10,000 m^2, its capacity is 20,000 m^3, and its actual filled volume is 13,000 m^3. Radionuclides with total activity of 5.2×10^{13} Bq are collected in the basin. Annually, 1,300 m^3 of LRW are sent to the basin from the chemical and metallurgical plant, and the same volume of LRW is sent to the RKh-1 sludge-storage facility after some retention.

Sludge-Storage Facilities

The RKh-1 is an on-surface LRW storage facility with a 0.45-m-thick loam wall at the bottom and along the slopes. The wall is covered with an 0.5-m-thick soil layer. Its area is 20,000 m^3, its capacity is 100,000 m^3, and its actual filled volume is 70,000 m^3. The total radionuclide activity is 6.66×10^{13} Bq. Annually, up to 100,000 m^3 of LRW come to the sludge-storage facility (which is used as an intermediate settler) from the enrichment and UF_6 production plants, as well as from clean-up facilities. The same volume of solution is sent to the RKh-2 sludge-storage facility.

The RKh-2 is an on-surface storage facility with a 0.45-m-thick loam wall at its bottom and along the slopes. The wall is covered with an 0.5-m-thick soil layer. The sludge-storage area is 46,000 m^2, its design capacity is 210,000 m^3, and its actual filled volume is 84,500 m^3. The sludge-storage facility receives 100,000 m^3 of sludge annually from RKh-1, and, after settling, the same volume is sent to underground disposal. The radionuclide activity in the sludge-storage facility is estimated at 3.33×10^{13} Bq.

Water-Storage Reservoirs

VKh-1 is a water-storage reservoir used for settling and interim storage of wastewaters. Its bottom has no waterproofing. Its area is 300,000 m^2, its volume is 500,000 m^3, and its actual filled volume is 375,000 m^3. The collected radionuclides activity is 2.96×10^{12} Bq.

VKh-3 is a water-storage reservoir that is part of the SCC's complex of clean-up facilities, and is used for balancing, settling, and interim storage of special sewage waters before cleaning. The reservoir's bottom is a natural soil, its area is 1.362 million m^2, its volume is 2.47 million m^3, and its actual filled volume is 2.336 million m^3. The total radionuclides activity is 2.22×10^{14} Bq. Annually, about 1.8 million m^3 of LRW enter the water reservoir from production works, and the same volume of wastewaters is sent to VKh-4.

VKh-4 is a water-storage reservoir placed at the cascade after the VKh-3, before the clean-up facilities. Its bottom has no damp proofing. Its area is 1.35 million m^2, its volume is 4.35 million m^3, and its actual filled volume is 2.88 million m^3. The total radionuclide activity is 4.44×10^{12} Bq. Annually 1.8 million m^3 of wastewater enter the water reservoir, and the same volume is sent to the SCC's clean-up facilities.

Underground Disposal of LRWs

Site 18 is used for low-level waste disposal (up to 6,300 m^3/day), and site 18a is used for medium-level waste disposal (up to 550 m^3/day). The latter site was also used for experimental disposal of limited portions of wastes, which were conditionally considered high-level wastes. Currently, the high-level wastes are not disposed. The underground storage sites are situated close to the main production complexes of the SCC within the limits of its SPZ. The wastes contain uranium fission products, including isotopes of strontium, zirconium, niobium, ruthenium, cesium, cerium, etc., as well as nonrecoverable microconcentrations of uranium and transuranium elements, salts, detergents, acids, alkalines, and finely dispersed solid materials.

Before disposal, treatment of the wastes is carried out at the SCC's clean-up facilities and at the radiochemical plant in order to prepare the wastes for the geological medium. Also, an additional extraction of long-lived transuranium elements is carried out before disposal.

The wastes are transferred to disposal sites through pipelines. Pumping is carried out through a system of injection wells (there are 37 wells in total). The injection pressure is as high as 2 MPa at site 18 and as high as 1.2 MPa at site 18a. Within and outside the limits of the disposal sites, 244 observation wells are used for determining the geological medium state and waste migration control.

The medium-level waste disposal at site 18a is carried out into the second sandy stratum, lying at a depth of 315–340 m. The strata used for LRW disposal are separated, spread, and covered by weakly permeable loam floors, isolating the strata containing the LRW from the surface of shallow underground waters. The natural speed of groundwater movement in the strata is 3–5 m a year. Radionuclides are intensively retained by the rocks.

Special geological prospecting work and investigations preceded creation of the underground disposal sites. These investigations substantiated, in principle, the disposal feasibility and safety. The license for underground disposal has been received.

SRW Storage Facilities

Over the more than 40-year period of the SCC's operation, more than 130,000 tons of SRWs have been generated. Data on accumulated SRWs and LRWs and characteristics of the wastes' storage and disposal facilities are presented in *Table 7.4*.

Radiation Situation in the SCC Area

The radiation levels in the SCC area are estimated by comparing radiation monitoring data of the environment around the SCC with the sanitary standards and, if necessary, with background levels.

Radiation monitoring is carried out in the observation zone (its area is 1,560 km^2 and its radius is 15–20 km) of the following objects:

- Releases into the atmosphere (19 radionuclides).
- Wastewaters (17 radionuclides).
- Near-surface layer of the atmospheric air at nine stationary points, located in the town and in a 15–50 km radius around the SCC (10 radionuclides and gamma background levels).

Table 7.4. Amount of accumulated SRWs and LRWs, and characteristics of their storage sites at the SCC.

Description of storage site	Period of operation		Volume, 1,000 m^3		Area, $10^3 m^2$	Amount of RW, $10^3 t$	RW activity		Notes
	Start	End	Design	Actual			Specific, Bq/l	Total, Bq	
SRWs									
Earthen trench-type burials without damp-proofing of bottom and slopes. As far as the burial is filled, vertical leveling with soil on waste top is carried out. 17 burials.	1955–1970	1987	166.8	146.7	53.66	84.41		–	SRW of the uranium enrichment plant, UF_6 production, chemical and metallurgical and radiochemical plants, production reactors, containing ^{234}U, ^{235}U, ^{238}U, ^{60}Co, ^{65}Zn, ^{137}Cs, ^{90}Sr, ^{103}Ru, ^{106}Ru; ^{239}Pu radionuclides
SRW storage facilities. Underground reinforced concrete structures with reinforced concrete cover. External and internal isolation with bitumen, concrete (or asphalt) laid on structure's bottom. In some cases they are lined with stainless steel.	1955–1992		129.2	95.35	18.04	46.7		1.1×10^{15}	SRW containing ^{60}Co, ^{65}Zn, ^{14}C, ^{137}Cs, ^{90}Sr, ^{95}Zr, ^{95}Nb, ^{103}Ru, ^{106}Ru; ^{137}Cs; ^{234}U, ^{235}U; ^{239}Pu radionuclides
SRWs in total			–	–	–	131.2		$>1.1\times10^{15}$	
LRWs									
Basins B-1, 2, 25. On-surface type storage facilities. Loamy isolation layer on bottom and slopes with 1-m-thick soil layer above loamy shield.	1961–1965	1982	305	187	122	186.75		4.6×10^{18}	Medium-level wastes, containing ^{90}Sr; ^{137}Cs; ^{144}Ce; ^{60}Co, ^{106}Ru; ^{235}U, ^{238}U; ^{239}Pu radionuclides
On-surface-type sludge storage facilities PKh-1,2. Loamy isolation layer on bottom and slopes with soil layer above loamy shield.	1961–1971		310	155	66	155		1.0×10^{14}	Medium-level wastes, containing ^{90}Sr; ^{137}Cs; ^{144}Ce; ^{103}Ru; ^{238}U, etc., radionuclides
Water storage reservoirs VKh-1, 2, 3, flowing, used for settling and intermediate holding of wastewaters. Bottom sediments are accumulated in the water reservoirs.	1955–1960		7,320	5,580	3,012	5,580		2.3×10^{14}	Low-level wastes, containing ^{90}Sr; ^{90}Y; ^{95}Zr; ^{95}Nb; ^{137}Cs; ^{144}Ce; ^{103}Ru, ^{106}Ru radionuclides
Underground LRW disposal sites (18, 18a)	1963		65,800	37,300	7,000	40,000	1.85×10^{3}–2.6×10^{11}	2.06×10^{19}	Low- and medium-level wastes, containing ^{90}Sr; ^{95}Zr; ^{95}Nb; ^{137}Cs; ^{144}Ce; ^{106}Ru; ^{238}U; ^{239}Pu radionuclides
LRWs in total			–	–	–	–	–	2.52×10^{19}	

- Soil, grass, and snow in the town and in 20 settlements situated around the SCC in a 15–20 km radius (^{90}Sr and ^{137}Cs).
- Groundwaters in 77 wells at the underground disposal site territory and in 48 wells outside the site.
- Open water reservoirs at the Tom and Ob Rivers. Monitoring is carried out twice a year 150–200 km downstream of the SCC's wastewater outlet, and once a year 700–800 km downstream of the outlet. Periodically, in summer, lakes are examined in the SCC's area.

Results of monitoring in 1993 and 1994 show that radioactive products released into the atmosphere were in the range of 0.1–0.6% of the maximum permissible levels in the areas surrounding the SCC (including the sum of alpha-emitting nuclides, 0.3%; the sum of beta-emitting nuclides and noble radioactive gases, from 0.1–0.3%; ^{131}I, from 0.2–0.4%; ^{90}Sr, 0.6%). In 1994, the absolute values of radionuclide content in the near-surface layer of the atmosphere in the area surrounding the SCC were as follows ($\times 10^{-7}$ Bq/l):

	Seversk	Tomsk	Pobeda settlement (background level)
^{239}Pu	13–21	0.3–2.0	9–16
^{90}Sr	0.01–0.03	0.037	0.037–0.07
^{137}Cs	0.037–2.6	0.037–0.11	0.037–0.11

The content of ^{103}Ru, ^{106}Ru, ^{144}Ce, ^{95}Zr, and ^{95}Nb was in the range of $(0.022–0.074) \times 10^{-7}$ Bq/l in all the marked points. The ^{131}I content was 1.11×10^{-5} Bq/l in all the marked points.

Radionuclides released with the SCC's wastewaters at the point of their discharge into the Tom River were in the range of 0.6–76.5% of the specified maximum permissible levels. At the settlements of Chernilshchikovo and Samus, situated along the Tom River downstream of the discharge point, the ^{54}Mn, ^{56}Mn, ^{58}Co, ^{59}Fe, ^{65}Zn, ^{89}Sr, and ^{144}Ce radionuclides were not detected even with lower sensitivity limits of measurement of two to five orders of magnitude less than the corresponding PC_B values.

With lower sensitivity limits of measurement, the following values were detected: 0.67 Bq/l for ^{137}Cs, 0.19 Bq/l for ^{90}Sr, 0.52 Bq/l for ^{106}Ru, and 4.1×10^2 Bq/l for tritium (measurements were 100–1,000 times less than the corresponding PC_B values). Radionuclides were not detected in groundwaters in observation wells of the local monitoring system (within a 12–15-km radius). The above-mentioned nuclides were not detected in samples of artesian drinking water.

According to the environmental monitoring data for 1995, the radiation level in the area surrounding the SCC can be specified as follows: Radioactive releases into the atmosphere did not exceed 0.2–0.3% of the maximum permissible values; radionuclide releases at the point of the LRW outlet into the Tom River were in the range of 0.7–33% of the maximum permissible level.

Radionuclide content in the near-surface layer of the atmosphere was close to background values – that is, by one to eight orders of magnitude lower than the corresponding permissible concentration values as specified by the NRB-76/87 rules for atmospheric air.

^{137}Cs and ^{90}Sr content in the observation zone soil is within the limits, and corresponds to the global level of $(1.11–3.7)\times10^9$ Bq/km^2. The only exception is the territory situated at the leeward side (to the northeast) 8–13 km from the SCC's industrial area, where ^{137}Cs content is two to three times higher than the global level; this is a result of many years of operation of the SCC production works. ^{137}Cs, ^{106}Ru, ^{90}Sr, and tritium radionuclides were not detected in groundwater samples of the local observation wells system, even though the sensitivity limits of measurement were 100–1,000 times lower than the permissible concentration levels set by the NRB-76/87 rules for drinking water.

The release of radioactive products that occurred during the accident on 6 April 1993 had no significant effect on radiation levels. The contamination was local and was mainly from ^{103}Ru, ^{106}Ru, ^{95}Zr, and ^{95}Nb radionuclides. At present, the gamma-exposure rate on the trace of the release does not exceed 20 μR/hr. In Georgievka village, the exposure rate has decreased to the natural background level (6–8 μR/hr).

In 1996, the same number of irradiated uranium slugs of operating uranium-graphite production reactors were reprocessed as were processed in 1995, and the radionuclide releases and discharges were at the 1995 level.

Emergency Releases of Radioactive Products

During the period of the SCC operations, 36 radiation accidents have occurred. The most serious accident (class 3 event, according to the International Nuclear Events Scale) took place on 6 April 1993.

With regard to the other 35 emergency situations, two radiation accidents should be noted. In June 1977, while reprocessing irradiated uranium slugs at the radiochemical plant, some insufficiently cooled slugs got into the processed lot. That led to elevated releases of ^{131}I (8.4×10^{11} Bq) into the atmosphere for two weeks. The monthly maximum permissible level was exceeded by 2.4 times. On 11 May 1977, an 8–14-m^3 spread of LRWs occurred at the radiochemical plant as a result of a blowhole in a pressure pipeline at an injection well while pumping LRWs. This led to contamination of a 200–220-m by 10–40-m plot. The gamma-exposure rate reached 140 R/hr. Later, the entire contaminated zone was cleaned up.

Accident of 6 April 1993

On 6 April 1993 at 12:58 (local time), an explosion occurred in a process apparatus (AD-6102/2) at the radiochemical plant of the SCC. The blast was due to a rapid increase in pressure in the process apparatus while preparing a uranium solution for an extraction purification. The explosion and release of the gas and aerosol mixture led to a breakdown of the upper ceilings of the chamber, where the apparatus was installed, as well as destruction of part of the shop building's structure, igniting a section of the roofing (the fire was contained in 10 minutes) and releasing some radioactive products contained in the gas and aerosol mixture into the atmosphere. The apparatus was a stainless steel vessel; its volume was 34.15 m^3. At the time of the explosion the vessel contained 25 m^3 of uranium nitric acid solution, which at the preceding stage was in contact with an organic solution (30% of tributhyl-phosphate in kerosene). The solution contained 8,773 kg of uranium and 320 g of plutonium; its total activity was 2.07×10^{13} Bq, including 7.4×10^{11} Bq of alpha-emitting nuclides. Two settlements are located in the contamination zone: Georgievka (16 km from the radiochemical plant; 30 inhabitants; gamma-exposure rate from 18–45 μR/hr) and Chernaya Rechka (34 km from the radiochemical plant; gamma-exposure rate from 12–50 μR/hr).

Table 7.5. Contaminated lands surrounding the SCC.

Distribution of the contaminated land areas by exposure rate level, μR/hr	Contaminated lands area, km^2			
	Including the territories of			
	Production site	SPZ	Observation zone	Total
Up to 60	3.838	–	–	3.838
60–120	1.558	–	–	1.558
120–240	0.958	0.30	–	1.258
240–1,000	1.697	–	–	1.697
>1,000	2.041	–	–	2.041
Total	10.092	0.30	–	10.392

The remaining recorded accidents at the SCC have not led to releases of radioactive products into the environment that exceed the specified maximum permissible levels. The total area of contaminated lands was 10.392 km^2 on 1 January 1996. The data on the contaminated lands are presented in *Table 7.5*. The soils are contaminated with ^{137}Cs and ^{90}Sr radionuclides. There are no published data on ^{90}Sr and ^{137}Cs distribution on the contaminated lands. More than 9.0 km^2 of the total contaminated land area at the SCC are located in the underground LRW disposal site territories, as well as in the B-1, B-2, and B-25 basins; RKh-1 and RKh-2 sludge-storage facilities; and VKh-1 and VKh-2 water-storage reservoirs.

The efforts of the SCC to minimize radionuclides' effects on the population and the environment were to some degree successful. At the initial stage of the radiochemical plant (the main source of LRW) operation, minimum releases were ensured due to the presence of basins for storage of medium-level process wastes. Putting underground LRW disposal sites into operation enabled the localized storage of hundreds of millions of curies of long-lived radionuclides within the limits of the SPZ and allotment area to be localized. The SRWs are isolated in reinforced concrete facilities or in suitable burials.

On the whole, it can be concluded that the stable operation of the enterprise has ensured a satisfactory radiation situation both at the production site and in the dwelling areas of the SCC's impact zone.

7.2.3 The MCC

The plans concerning the construction of the MCC were made in 1950. At that time, it was decided to place the main objects of the MCC in a rock mass on the right bank of the Yenisei River, 60 km downstream of the city of Krasnoyarsk. The MCC's underground site was chosen at the place where the Atamanov chain of hills meet the Yenisei River. The territory for the town's construction was chosen 10 km to the south of the main production site on the former flood plain of the Yenisei River.

The MCC's territory is about 360 km^2 along the right bank of the Yenisei River. Southwesterly winds predominate. The observation zone dimensions are: 15 km toward the predominating winds bearing to the northeast, 9 km of the windward side from the southwest, and 2–8 km toward other directions. Settlements in the area controlled by the MCC include the town of Zheleznogorsk, with a population of more than 80,000; about 600 people in the village of Dodonovo; about 400 people in the village of Bolshoi Balchug; and about 5,000 people in the Atamanovo settlement.

The river flowing close to the MCC's territory is a controlled non-freezing water stream from the hydroelectric power plant.

The MCC was designated for weapons-grade plutonium production. The plutonium is produced in uranium-graphite reactors. The irradiated uranium is processed at the radiochemical plant to separate uranium and plutonium and decontaminate them from fission products. Plutonium dioxide and fused uranyl-nitrate are the MCC's end products, which are sent to other enterprises of the Ministry of the Russian Federation for Atomic Energy (Minatom). Liquid releases from the reprocessing, as well as gaseous effluents, undergo decontamination at the clean-up facilities of the MCC. SRWs are sent to special storage facilities of the enterprise.

The plant's design called for it to include a reactor and radiochemical production works, a nuclear power and heating facility, and water supply and ventilation facilities located deep within the nearby mountain range. In addition, the project called for construction of components for LRW and SRW storage and processing, for heat supply, and for other auxiliary services necessary for normal operations of the industrial enterprise and for the population.

Three production reactors have operated at the MCC. The first of them, AD, was started up on 26 August 1958; the second, ADE-1, was started up on 20 July 1961. Both are thermal, uranium-graphite, water-cooled channel-type reactors. Natural uranium slugs were loaded into the core. The reactors' safe operation at a set power level and reliable halting of the chain nuclear reaction in the core in case of an emergency were ensured by control, monitoring, and protection systems. The reactors worked in single-pass mode, disposing cooling water into the Yenisei River. The AD and ADE-1 reactors were shut down on 30 June 1992 and on 29 September 1992, respectively.

The third, a two-purpose ADE-2 reactor, was put into operation in 1964. It is used for both plutonium and electric power production, as well as for water heating (since 1966) for the town of Zheleznogorsk. The power-and-heating complex includes six turbo generators, 30 steam generators, a circuit pumping plant, an electric substation, and other facilities.

The radiochemical plant was put into operation in 1964 and is designated for the extraction of plutonium from irradiated uranium. The plant's basic process flowchart includes dissolution of metallic uranium in nitric acid, multistage extraction reprocessing of the solution to separate uranium and plutonium and to decontaminate them from radioactive fission products, and deep purification of the plutonium concentrate by the sorption method. Uranyl-nitrate and plutonium dioxide are the end products of the reprocessing of the irradiated uranium slugs. Because of a sharp reduction in weapons-grade plutonium production, the reprocessing of irradiated uranium slugs has ceased at the plant. High-level LRWs are sent to special storage facilities. Medium- and low-level RWs, after collecting and corresponding treatment, are sent to the Severnyi site for underground disposal into deep geological strata-collectors. Gaseous and aerosol effluents undergo multistage cleaning and then are discharged into the atmosphere.

In 1977, the construction of the RT-2 plant was started. The plant is designated for the interim storage and reprocessing of SNF from former Soviet NPPs using VVER-1000 nuclear reactors. The unburnt uranium and generated plutonium from these plants can be returned to the nuclear fuel cycle for repeated use.

The RT-2's design capacity is 1,500 tons of irradiated uranium a year. Mixed uranium-plutonium oxide (MOX-fuel), fuel assemblies, and regenerated uranium are the plant's end products. The RT-2 plant process flowchart is based on extraction technology of uranium and plutonium separation and purification. At the beginning, the ultimate disposal of the LRWs generated by the RT-2 plant was to be at a new deep well injection area west of the Yenisei River. This approach has been abandoned because of the concerns raised by transportation of the wastes to this site via pipeline in a tunnel under the Yenisei River. The final design of the

RT-2 plant stipulates that LRWs, emerging from SNF reprocessing, must undergo fractionation followed by solidification into glass- or mineral-like compositions. The RW management processes, adopted in the plant's design, are in agreement with present-day concepts, and allow the maximum decrease of RW volumes, and provide safe storage conditions. The SNF storehouse design capacity is 6,000 tons. At present it is filled to 16%. After 1985, there was a sharp reduction in funding of further construction of the RT-2, and in 1990, the construction was completely stopped due to a lack of funding.

Overall Characterization of Accumulated RWs at the Enterprise

The whole nuclear production complex, including three reactors and a radiochemical plant, is situated underground (250–300 m deep) and, thus, unlike the IA Mayak and the SCC, has reliable biological shielding. The complex is equipped with a ventilation system with filters that ensures protection against activity from the outside. As a result of the MCC operation, large amounts of high-, medium-, and low-level LRWs and SRWs have been collected. The SRWs and LRWs are kept in storage facilities on the MCC's territory.

LRWs

LRWs resulting from production operations are sent, depending on the activity level, either to cleaning facilities or collected in special tanks or in open storage reservoirs. After corresponding treatment and cleaning they are sent to an underground disposal site (Severnyi). Decontaminated waters are discharged into the Yenisei River.

Basin 365 is an open-type, water-storage reservoir, constructed on the first large flood-plain terrace of the Yenisei River. It is designated to receive and temporarily store the reactors' emergency waters and off-grade nonprocessed wastewaters of the radiochemical plant before sending them to cleaning facilities. LRWs are isolated from groundwaters with an anti-filtering shield of clay, two asphalt layers on the bottom and slopes, as well as a bottom and bank draining system for interception and collection of leakages in case of damage to the waterproofing.

Basin 366 is an open-pit-type, water-storage reservoir, constructed on the first large flood-plain terrace of the Yenisei River. It was built by hydraulic deposition of soil and is designated to receive decontaminated (in accordance with set standards) waters from the MCC's cleaning facilities for holding, settling, and filtration before their discharge first into a stream and then into the Yenisei River. During operation, the water filtrates through the bottom and the dam body and, in case of excessive filling, it discharges over the berm.

Basin 354a is an open-pit-type, water-storage reservoir, built in practically waterproof rocks and designated to receive, balance the composition, and temporarily store regeneration solutions and sludges from cleaning facilities, as well as low-level wastes and condensate after evaporation of the radiochemical plant process wastes, before sending them to an underground disposal site. The wastes are isolated from contact with groundwaters through a two-layer antifiltering shield on the bottom and slopes and through a drainage system between the shield's layers, as well as through the presence of a thick covering of uniform and practically waterproof clays.

Basin 354 is 100 m from basin 354a, on a site with identical engineered geological and hydrogeological conditions. Its design is analogous to that of basin 354a. At present, the basin is completely empty and work on its elimination is being conducted.

The Underground LRW Disposal Site

The Severnyi site is used for disposal of low-level RWs (up to 800 m^3/day) into the second sandy stratum, and medium-level RWs (up to 500 m^3/day) into the first sandy stratum. Earlier experimental disposal of some portions of wastes, conditionally attributed to high-level ones, was carried out in the first sandy stratum.

The site is situated 12 km from the MCC's main production works within the limits of the SPZ of the enterprise. The disposed wastes contain radionuclides of uranium fission products, including strontium, cesium, zirconium, niobium, ruthenium, cerium, and trace amounts of uranium and transuranium elements that are unrecoverable in the solutions' processing. Before underground disposal, RW treatment is carried out at the MCC's cleaning facilities and at the radiochemical plant to provide compatibility of the wastes with the geological medium, as well as additional recovery of long-lived transuranium elements.

The wastes are transferred by pipelines. The pumping is carried out through a system of 16 injection wells, using an injection pressure of about

2 MPa. Within and outside the disposal site, 70 observation wells are used to control the state of the geological medium and the wastes' migration.

The first and second sandy strata used for disposal are at 300–500 m deep and 180–280 m deep, respectively. The strata are underspread, separated, and covered by loam floors, isolating the strata containing RWs from surface and shallow underground waters. The natural speed of water movement is 5–6 m/yr in the first stratum, and 10–15 m/yr in the second stratum. Radionuclides are intensively retained by the rocks.

Special geological prospecting works and explorations preceded creation of the Severnyi underground disposal site. The work substantiated the feasibility of RW injection and the safety of disposal. The disposal is licensed.

The underground disposal of LRWs eliminated the impact of huge amounts of radionuclides on the population and on the environment. The state of the wastes and the geological formations used to contain them are controlled by conducting complex observations and measurements. The results measure the disposal sites' safety and indicate whether additional isolation of RWs is necessary. Probable complications and emergency situations while carrying out the disposal will not lead to disastrous consequences connected with large-scale effects on the environmental systems.

RW Storage Facilities

Great amounts of SRWs and LRWs have been generated during the years of operations at the MCC. Data on the accumulated wastes are presented in *Table 7.6*.

Radiation Situation in the MCC Area

Radionuclide releases into the atmosphere in 1993 did not exceed the standard levels by any components. The releases of radioactive noble gases were 0.6% of the maximum permissible value. The sum of released beta-emitting nuclides was 0.9% of the standard value. The mean annual volumetric activity of radionuclides in the near-surface layer of the atmosphere in the SPZ of the MCC was lower than PC_B for ^{60}Co by 4.3×10^6 times; for ^{106}Ru by 3.8×10^5 times; for ^{144}Ce by 2.8×10^5 times; and for ^{137}Cs by 3.0×10^6 times.

Since decommissioning of the AD and ADE-1 single-pass reactors, the volume activity level in the near-surface layer of the atmosphere has

Table 7.6. Amount of accumulated SRWs and LRWs, and characteristics of their storage sites at the MCC.

Description of storage site	Period of operation		Volume, 1,000 m^3		Area, 10^3m^2	Amount of RW, 10^3t	RW activity		Notes
	Start	End	Design	Actual			Specific, Bq/l	Total, Bq	
SRWs									
Four SRW storage facilities. Reinforced concrete reservoirs, deepened into the ground. Bottom is a layer of compacted crushed rock 70-mm thick impregnated with bitumen and covered by a 35-mm asphalt layer.	1963		27.4	24.2	5.0				Fine granular SRW of *Groups II* and *III* in shielded containers. Large-sized SRW of *Groups II* and *III*.
Seven SRW storage facilities. Filled earthen trenches in compacted loam. As soon as the trenches are filled, they are covered by a 1-m soil layer.	1963	1983	111.6	109.6	38.3				*Group I* SRW burial.
Four SRW storage facilities. Reinforced concrete shafts lined by stainless steel.	1961–1964		1.86	1.7	0.06				*Groups II* and *III* reactor SRW.
SRWs in total including:						105.17			SRW contain: ^{60}Co,
Group I with exposure rates in the range of 0.015–5.5 mR/s.						52.17			^{90}Sr, ^{95}Zr, ^{95}, ^{103}Ru, ^{106}Ru, ^{137}Cs
Groups II and *III* with exposure rates in the range of 5.5–250 and more mR/s.						53			^{141}Ce, ^{144}Ce, ^{238}U, ^{239}Pu, etc.

Notes:
The SRW groups include:
Group I: household rubbish, contaminated work clothes and footwear, breathing apparatuses, package materials, cleaning materials, wooden containers, wastes from repair shops, dismantled equipment, tubing scrap, building refuse, etc.
Group II: graphite bushing, fuel-channel briquettes, deteriorated metallic components, radiochemical laboratories' glassware, building refuse, filters, etc.
Group III: instrument sensors, wastes of repair and construction works at radioactively contaminated sites, radioactive material spreads and spills, radiochemical production works' SRW that contain alpha-emitting nuclides, etc.
The SRW gamma-exposure rate measurements are made immediately adjacent to a source.
Distribution of SRW while filling the storage facilities was carried out only on the basis of gamma-exposure rate measurements. No other parameters were measured.

Table 7.6. Continued.

Description of storage site	Period of operation		Volume, 1,000 m^3		Area, $10^3 m^2$	Amount of RW, $10^3 t$	RW activity		Notes
	Start	End	Design	Actual			Specific, Bq/l	Total, Bq	
LRWs									
High-level LRW storehouse. Complex includes 24 stainless steel 300-m^3 tanks, placed in canyons. Canyon walls are lined with stainless steel. Each canyon is covered by a 1-m concrete plate. The tanks are provided with coil coolers.	1963–1973		6.84	2.02	4.4	2.02	Up to 1.85×10^{13}	3.07×10^{18}	The high-level solutions are received for storage and processing from the radiochemical plant. They contain: ^{238}U, ^{239}Pu, ^{95}Zr, ^{95}, ^{103}Ru, ^{106}Ru, ^{144}Ce, ^{137}Cs, ^{90}Sr, etc.
Medium-level LRW storehouse. Complex includes reinforced concrete tanks, nine of them have 3,000-m^3 capacity, eight have 8,500-m^3 capacity. The tanks are lined with stainless steel or with carbon steel with epoxy coating. The tanks are equipped with ventilation systems for solutions' cooling.	1964–1965		94.55	53.1	4.0	53		0.81×10^{18}	The liquid medium-level wastes are received from the radiochemical plant. They contain: ^{238}U, ^{239}Pu, ^{95}Zr, ^{95}, ^{103}Ru, ^{106}Ru, ^{144}Ce, ^{137}Cs, ^{90}Sr, etc.
Medium-level pearlite sludge storage facility. A stainless steel tank placed in a compartment with reinforced concrete walls lined with stainless steel.	1986		0.5	0.17	0.078	0.17			The sludge contains 50 m^3 of solids radionuclides from process solutions.
Storage basins	1958–1966		794	520	130	566.8		1.4×10^{15}	The solution and bottom sediments contain: ^{137}Cs, ^{144}Ce, ^{152}Eu, ^{154}Eu, ^{60}Co, ^{106}Ru
Underground LRW disposal site	1967		11,000	5,000	6,300	5,000	1.85×10^{5}–1.48×10^{11}	1.07×10^{19}	^{90}Sr, ^{95}Zr, ^{95}, ^{103}Ru, ^{106}Ru, ^{137}Cs, ^{144}Ce, ^{235}U, ^{238}U, ^{239}Pu radionuclides are contained.
LRWs in total			11,895.9	5,575.3	6,438.5	5,622		1.46×10^{19}	

fallen eightfold. In the settlements nearest to the MCC (the Bolshoi Balchug village and the town of Zheleznogorsk), ^{137}Cs is primarily detected in the near-surface layer of the atmosphere. Its volume activity is 7.6 times lower than the PC_B. On the whole, the effect of gaseous and aerosol effluents from the MCC production works on the contamination in the SPZ and in the observation zone is practically not detectable against the global background level.

Since the shutdown of the AD and ADE-1's single-pass reactors, the release of radionuclides into the Yenisei River has mainly been determined by short-lived isotopes (^{24}Na, ^{32}P) in the cooling water of the control and protection system of the two-purpose ADE-2 reactor. The volume activity of the water discharged into the Yenisei River is 1.2–7.0 PC_B for ^{24}Na, and 0.05–1.5 PC_B for ^{32}P. On the whole, the total release of all radionuclides did not exceed the standards and was in the limits of 0.3–6.0% of calculated values of the maximum permissible release. The volume activity of radionuclides in the river's water does not exceed 0.3 PC_B at the discharge location, 0.08 PC_B at 500 m from the discharge location downstream, and 0.015 PC_B at 15 km from the discharge location downstream (1 km upstream of the first settlement on the right bank of the Yenisei River, the Bolshoi Balchug village). The values of ^{239}Pu and ^{240}Pu volume activity (totaled) are lower than the sensitivity limit of the measurement method, and they do not exceed 8.0×10^{-5} PC_B. The maximum values of ^{90}Sr and ^{137}Cs volumetric activity are 1.2×10^{-3} and 6.0×10^{-3} of PC_B, respectively. If the Yenisei River were used as the sole water supply for the first settlement on the right bank (the Bolshoi Balchug village), the effective dose limit due to the river water consumption would be 5 μSv/yr.

On the whole, after decommissioning of the single-pass reactors, the water surface exposure rate and volumetric activity of all radionuclides (totaled) in the water do not exceed standard values set by the NRB-76/87 at the discharge location. The radioecological situation in the flood plain of the Yenisei River is mainly determined by the formerly conducted discharge of the single-pass AD and ADE-1 reactors' cooling waters. The values of the exposure rate do not exceed 10–15 μR/hr along the river, 15–500 km downstream of the MCC discharge location where populations are located. On particular islands and in some local sections of the flood plain 15–250 km downstream of the MCC discharge location, there are "spots" with exposure rate levels reaching 200 μR/hr.

Table 7.7. Contaminated lands at the MCC.

Distribution of the contaminated land areas by exposure rate level, μR/hr	Contaminated lands area, km^2			
	Including the territories of			
	Production site	SPZ	Observation zone	Total
Up to 60	0.005	0.666	0.106	0.777
60–120	–	0.149	–	0.149
120–240	3.297	0.060	3.394	6.751
240–1,000	–	0.050	–	0.050
>1,000	–	0.062	–	0.062
Total	3.302	0.987	3.500	7.789

In the 300-km zone, the radioactive contamination of the flood plain of the Yenisei River was caused mainly by two intense floodings in 1966 and 1988 that dispersed bottom sediments, containing radionuclides, to islands and flood-plain areas. At present in that zone there are some areas with gamma-exposure rates ranging from 30–200 μR/hr.

As of 1 January 1996, 7.79 km^2 were contaminated. The lands are contaminated with ^{137}Cs and ^{90}Sr radionuclides. The data on the contaminated lands are presented in *Table 7.7.*

More than 5.7 km^2 of the total contaminated lands area at the MCC are in the underground LRW disposal site territory, as well as at basins 354, 354a, 365, and 366.

Reliable information about the situation in radioactively contaminated territories is needed to make decisions concerning remediation measures. For this reason, permanent observation and monitoring should be conducted, and a system of radioecological control of the areas and of radioactive contamination should be created.

7.3 Characteristics of Available Data

The data on the three industrial enterprises presented in this chapter give a rather complete representation of quantitative characteristics concerning accumulated RWs and their activity. Some years ago (on 1 January 1990 and on 1 January 1993), inventories of sites and objects for treatment, utilization, collection, storage, and disposal of RMs and ionized radiation sources were conducted.

The enterprises annually provide Minatom with data on the discharge of radioactive products into water reservoirs and effluents into the atmosphere, and on radioactively contaminated and remediated territories. Databases on discharges of radionuclides into water reservoirs and effluents into the atmosphere, as well as on radioactively contaminated lands, have been available since 1 January 1993.

Bibliography

Akleev, A.V., *et al.*, 1991, *Radioactive Contamination of the Environment in the South Urals Region and Its Effect on the Health of the Population,* Academician L.A. Buldakov, ed., TsNIIAtominform, Moscow, Russia.

Fetisov, V.I., 1993, The IA Mayak in front of a new challenge: The conversion, *Atom-pressa*, No. 31.

Glagolenko, Yu.V., Dzekun, E.G., Rovnyi, S.I., and Ufimtsev, V.P., 1996, The reprocessing of spent nuclear fuel at the IA Mayak, In: *Abstracts of reports from the Third International Radioecological Conference "The Destiny of Spent Nuclear Fuel: Problems and Reality,"* 22–27 June, Krasnoyarsk, Russia.

IA (Industrial Association) Mayak, 1992, The environmental consequences of the IA Mayak activities, *The Information Sheet of the Environmental Center of the IA Mayak*, IA Mayak, Mayak, Chelyabinsk region, Russia.

IIASA (International Institute for Applied Systems Analysis), 1996, *Draft Final Report on Mayak Case Study*, IIASA, Laxenburg, Austria.

Kedrovskii, O.L., Rybalchenko, A.I., Pimenov, M.K., *et al.*, 1991, The deep burial of liquid radioactive wastes into porous geological formations, *Atomic Energy*, **70**(5):298–303.

Kruglov, A.K., 1994a, *How the Nuclear Industry was Created in the USSR*, TsNIIAtominform, Moscow, Russia.

Kruglov, A.K., 1994b, On the first heavy-water nuclear reactors in the country, In: *Information Bulletin of the Public Information Centre on Nuclear Energy*, No. 7–8, TsNIIAtominform, Moscow, Russia.

Kruglov, A.K., Reshetko, Yu.V., and Smirnov, Yu.V., 1992, On the restoration of radioactively contaminated territories, *Information Bulletin of the Public Information Centre on Nuclear Energy*, No. 10.

Lebedev, V.A., 1994, The Mining and Chemical Combine: Its past, present day, and future, In: *Collection of papers presented at The Second International Radioecological Conference "After the Cold War: Disarmament, Conversion, and Security,"* 12–16 September, Krasnoyarsk, Russia.

Makhonko, K.P., Pavlova, L.N., Martynenko, V.P., *et al.*, 1996, The estimation of the Siberian Chemical Combine effect on radioactive contamination of the environment in 1989–1994, *Atomic Energy*, **80**(3).

Malyshkin, A.I., Belousov, G.P., and Bushuev, S.I., 1994, The radiation situation in the SCC area, *Paper presented at The Second International Radioecological Conference "After the Cold War: Disarmament, Conversion, and Security,"* 12–16 September, Krasnoyarsk, Russia.

Malyshkin, A.I., Belousov, G.P., and Bushuev, S.I., 1996, The radiation situation in the SCC area, *Paper presented at The Third International Radioecological Conference "The Destiny of Spent Nuclear Fuel: Problems and Reality,"* 22–27 June, Krasnoyarsk, Russia.

Menshchikov, V.F., 1994, *Results of Examination for Ensuring Radiation and Environmental Safety of the SCC and Adjacent Territories*, RF Security Council Working Group, Moscow, Russia.

Menshchikov, V.F., and Yakimets, V.N., 1994, Criteria of considering projects for spent nuclear fuel reprocessing are necessary, In: *Collection of reports of The Second International Radioecological Conference "After the Cold War: Disarmament, Conversion, and Security,"* 12–16 September, Krasnoyarsk, Russia.

Merkin, V.I., 1996, The first production nuclear reactor of the Soviet Union: Its erection and starting-up, In: *Collection of reports delivered at the Jubilee Scientific and Practical Conference of Minatom's "50 Years of the Nuclear Science and Engineering of Russia,"* Moscow, Russia.

Minatom (Ministry of the Russian Federation for Atomic Energy), 1995, *The Federal Goal-Setting Program for Radioactive Wastes and Spent Nuclear Materials Management, Their Utilization and Disposal, for 1996–2005*, Affirmed by Decree of the Government of Russian Federation from 23 October, No. 1030, Minatom, Moscow, Russia.

Morozov, P.V., 1994, *The Requisite Card*, V.V. Zhidkov, ed., Zheleznogorsk, Krasnoyarsk Land, Russia.

Polyakov, A.S., 1996, Irradiated nuclear materials and radioactive wastes management, In: *Collection of reports delivered at the Jubilee Scientific and Practical Conference of Minatom's "50 Years of the Nuclear Science and Engineering of Russia,"* Moscow, Russia.

Rybalchenko, A.I., Pimenov, M.K., Kostin, P.P., *et al.*, 1994, *The Deep Burial of Liquid Radioactive Wastes*, IzdAT, Moscow, Russia.

The Administration of the Chelyabinsk region, 1994, *Information of the Chelyabinsk Region Administration on Radioactive Contamination of Soils of the Northern Part of the Chelyabinsk Region, as of 1993*, The Emergency Situations Committee, Chelyabinsk, Chelyabinsk region, Russia.

VNIICHT (All-Russia Research Institute of Chemical Technology), 1994a, *Restoration of Territories Contaminated as a Result of Activities of Enterprises Producing Nuclear Weapons Materials* (draft of Federal program), Minatom (Ministry of the Russian Federation for Atomic Energy), Moscow, Russia.

VNIICHT (All-Russia Research Institute of Chemical Technology), 1994b, Cadastre of sites for storage and disposal of radioactive wastes, Minatom (Ministry of the Russian Federation for Atomic Energy), Moscow, Russia.

Vorobyov, G.V., Dmitriev, A.M., Dyakov, A.S., Yershov, Yu.I., Osanov, D.P., and Popova, L.V., 1994, *Collection Plutonium in Russia: Environmental, Economical, Political Aspects of the Problem*, Centre of the Environmental Policy of Russia and the Centre of Nuclear Ecology and Energy Policy of the Social and Environmental Union, Moscow, Russia.

Zakharkin, B.S., 1993, The elaboration of technology of spent nuclear fuel reprocessing, In: *Pages of the History of VNIINM: The Recollections of Workers*, Vol. 2, TsNIIAtominform, Moscow, Russia.

Zemlyanukhin, V.I., 1996, The creation of radiochemical technology of plutonium production, In: *Abstracts of reports from the international symposium "Science and Society: History of the Soviet Nuclear Project (1940s–1950s),"* 14–18 May, Dubna, Russia.

Zhidkov, V.V., 1994, The radioecological situation in the SCC area, *Paper presented at The Second International Radioecological Conference "After the Cold War: Disarmament, Conversion, and Security,"* 12–16 September, Krasnoyarsk, Russia.

Zhidkov, V.V., and Shishlov, A.E., 1996, The radioecological consequences of operation of the works on the weapon grade plutonium production at the MCC: The results of the latest studies, *Paper presented at The Third International Radioecological Conference "The Destiny of Spent Nuclear Fuel: Problems and Reality,"* 22–27 June, Krasnoyarsk, Russia.

Chapter 8

Nuclear Weapons Production

The main enterprises of the Russian Federation (RF) related to nuclear weapons production include:

- The Combine Elektrokhimpribor, Lesnoy, Sverdlovsk region.
- The Industrial Association (IA) Start, Zarechny, Penza region.
- The Instrument Engineering Plant, Trekhgorny, Chelyabinsk region.
- The Electromechanical Plant Avangard, Sarov, Nizhniy Novgorod region.

As of 1 January 1997, the total amount of solid radioactive waste (SRW) accumulated at the enterprises of nuclear weapons production complexes was 4,301 m^3. The total amount of liquid radioactive waste (LRW) was 2,546 m^3. The total activity of SRW and LRW was 5.96×10^6 Bq (wastewaters with radionuclide contents lower than the PC_B are not considered in the category of LRWs). The data on radioactive waste (RW) accumulation in separate enterprises are presented in *Table 8.1*. The main storage facilities are reinforced concrete tanks.

Radionuclides releases into the environment by the enterprises of the nuclear weapons production complex are insignificant (less than 1% of maximum permissible values). For the entire time period of operation of the enterprises, no incidents occurred that might lead to radioactive contamination of the environment.

Table 8.1. Accumulated RWs at nuclear weapons production enterprises of the RF, as of 1 January 1997.

Enterprise	SRW					LRW				
			Characteristics of storage facilities					Characteristics of storage facilities		
	Vol., m^3	Total actvitiy, Bq	Year operation began	Vol., m^3	Level of filling, %	Vol., m^3	Total actvitiy, Bq	Year operation began	Vol., m^3	Level of filling, %
Combine Elektrokhimpribor	1,053	1.1×10^6	1983	2,633	40	586	9.6×10^5	1983	690	85
IA Start	556	7.4×10^5	1982	696	80	–	–	–	–	–
Instrument Engineering Plant	1,728	1.1×10^6	1981	2,160	80	1,960	9.6×10^5	1981	2,450	80
Electromechanical Plant Avangard	964	1.1×10^6	1983	1,378	70	–	–	–	–	–

Table 8.2. Planned increase in volumes of RW storage facilities at nuclear weapons production enterprises through 2000.

Enterprise	Additional volumes of storage facilities, m^3	
	SRW	LRW
Combine Elektrokhimpribor	4,000	2,100
IA Start	700	–
Instrument Engineering Plant	2,800	1,500
Electromechanical Plant Avangard	2,700	–

In the future, enlargement of RW storage facilities is planned at all enterprises of the nuclear weapons production complex. The data on the planned increase of volumes of SRW and LRW storage facilities at the separate enterprises through 2000 are presented in *Table 8.2*.

Bibliography

All the data presented in this chapter are taken from the Ministry of the Russian Federation for Atomic Energy (Minatom) annual statistical reports.

Chapter 9

Nuclear Explosions

9.1 Introduction

Nuclear weapons tests and nuclear device explosions were carried out in the Soviet Union from August 1949 (the first nuclear test) until 24 October 1990 (the last nuclear test). The number of Soviet nuclear tests totals 715, and the total number of exploded nuclear charges and nuclear explosive devices is 969.

In a report from the Russian Federal Nuclear Center–the All-Russia Institute of Experimental Physics (RFNC–VNIIEF) on nuclear weapons tests and peaceful nuclear explosions (PNEs), the following classification of nuclear explosion (NE) purposes is given:

- NWR – nuclear weapons-related test (weapon development or modification).
- SAM – study of accidental models and emergencies.
- WIE – weapons effects test (studies of injurious effects of a NE on military and civilian systems and equipment).
- FMS – fundamental and methodical study of the phenomena of a NE.
- ME – military exercise with a nuclear detonation.
- IE – industrial underground PNE and testing of PNE technologies.
- TIC – testing of industrial nuclear charges for use in peaceful activities.

The following classification is adopted for the types of Soviet nuclear tests (explosions):

- A surface explosion is a nuclear test from the earth's surface or from a tower.[1] In terms of physical criteria relating to the radiation environmental effects, the category of surface explosions includes all nuclear tests with a scaled height of burst (HOB) of no more than 35 m/kt$^{1/3}$.[2]
- An air explosion[3] is a nuclear test in the atmosphere with a scaled HOB of no less than 100 m/kt$^{1/3}$ (under these conditions, the expanding fireball does not touch the ground surface). Within this category are high-altitude explosions, for which the fireball size is comparable with the characteristic size of the atmosphere inhomogeneity (~7 km), and which are recognized as a subcategory of air explosions; this subcategory also includes space explosions.
- An underwater explosion is a nuclear test where the explosive device was under the water's surface.
- An underground explosion is a nuclear test where the explosive device was under the earth's surface.

For nuclear weapons testing in the Soviet Union, two test sites were established: the Semipalatinsk test site (STS) in Kazakhstan, where the first nuclear device was tested on 29 August 1949; and according to the 31 July 1954 decree of the Government of the Soviet Union, the Northern test site at the arctic archipelago Novaya Zemlya (NTSNZ). The first nuclear test at NTSNZ was on 21 September 1955.

Military nuclear tests were also conducted at three other locations outside the territory of the two test sites:

- At the missile testing range (MTR) in the Astrakhan region, missiles were launched with nuclear and thermonuclear warheads in order to be tested in upper layers of the atmosphere and in outer space.
- On 14 September 1954, an air NE of 40 kt was carried out in the course of army exercises at the firing grounds of the Soviet Ministry of Defense near Totsk (Orenburg region).
- On 2 February 1956, near Aralsk (Kazakh SSR), a surface explosion was carried out with an energy release of 0.3 kt.

[1]This category includes the NE on 30 October 1954 (aerial bomb), which had a low height of burst (HOB).

[2]Kiloton of TNT.

[3]This category includes the nuclear detonations conducted on 26 October 1954 and 10 September 1956, which had a scaled HOB of ~80 m/kt$^{1/3}$.

In addition to the nuclear weapons tests, beginning on 15 January 1965, peaceful underground NEs were also conducted in the Soviet Union. During the program, 173 nuclear charges and devices were exploded in 156 nuclear tests. A total of 119 nuclear tests and PNEs were conducted outside the territory of the nuclear test sites.

9.2 Data on Nuclear Weapon Tests and Peaceful Nuclear Explosions, Conducted from 1949 to 1990

Tables 9.1–9.6 provide historical information on Soviet nuclear weapons tests. The tables are self-explanatory.

9.3 Characteristics of Accumulated Radionuclides Due to Nuclear Explosions on the Territory of the Former Soviet Union as of 1 January 1996

9.3.1 The radionuclide composition of radioactive fallout

Radionuclides with half-lives of more than four years that are produced by NEs may be classified into three groups: nuclear fission products (FPs), radionuclides induced by explosion neutrons (IRs), and transuranium radionuclides (TRs). Also attributable to the category of TRs are some components of nuclear explosives (e.g., ^{239}Pu) and heavy nuclei formed as a result of nuclear reactions in the course of the explosion.

Considering the long-lived FP group, including 20 radionuclides with half-lives from 4.96 years (^{155}Eu) to 1.3×10^{11} years (^{147}Sm), leads to the conclusion that in the explosions' products it is realistic to expect measurable contents of ^{90}Sr, ^{137}Cs, ^{151}Sm, and ^{155}Eu and, possibly, ^{79}Se, ^{93m}Nb, ^{121m}Sn, ^{126}Sn, ^{152}Eu, and ^{154}Eu.

There are two ways the radionuclides can form in the IR group: (1) as a result of activation of the construction elements of the explosion device by neutrons produced during the explosion, and (2) as a result of these neutrons interacting with soil matter (in surface and low-altitude explosions). The second category also includes instances where the explosion

Table 9.1. Soviet nuclear weapons tests and PNEs, by calendar year.

Year	Number	Year	Number
1949	1	1971	23
1950	0	1972	24
1951	2	1973	17
1952	0	1974	21
1953	5	1975	19
1954	10	1976	21
1955	6	1977	24
1956	9	1978	31
1957	16	1979	31
1958	34	1980	24
1959	0	1981	21
1960	0	1982	19
1961	59	1983	25
1962	79	1984	27
1963	0	1985	10
1964	9	1986	0
1965	14	1987	23
1966	18	1988	16
1967	17	1989	7
1968	17	1990	1
1969	19		
1970	16		

Table 9.2. Soviet nuclear weapons tests and PNEs, by location.

Semipalatinsk test site	456
Northern test site (Novaya Zemlya)	130
Total at nuclear test sites	586
Russian Federation	91
European part	59
Asian part	32
Ukraine	2
Kazakhstan	33
Uzbekistan	2
Turkmenia	1
Total outside nuclear test sites	129

Table 9.3. Soviet nuclear weapons tests and PNEs, by type.

Air explosions	177
Surface explosions	32
Underwater and water surface explosions	5
High-altitude explosions	1
Space explosions	4
Total number of atmospheric, underwater, and space explosions	219
Tunnel tests	245
Shaft tests	251
including criteria explosions	5
Total number of underground tests	496
Total number of tests	715

Table 9.4. Soviet nuclear tests and exploded nuclear devices, by purpose.

Purpose	Number of tests	Number of nuclear charges and nuclear devices
NWR	445	637
SAM	25	42
WIE	52	69
FMS	36	47
AT	1	1
Total for military purposes	559	796
IE	124	135
TIC	32	38
Total for peaceful purposes	156	173
Total number of nuclear tests and PNEs		715
Total number of nuclear charges and devices		969

Table 9.5. Total energy release of Soviet nuclear tests during certain time periods and in specific regions, kt (rounded values within 10 kt).

Time period	STS	NTSNZ	Industrial explosions and Azgir MEs	Kapustin Yar, Totsk, Aralsk regions	Total
1949–1962	6,580	239,630	0	1,030	247,240
1964–1975	2,940	23,770	590	0	27,300
1976–1990	7,900	1,940	1,010	0	10,850
Total	17,420	265,340	1,600	1,030	285,390

Table 9.6. Energy release of Soviet nuclear tests over the years and in specific regions, kt (rounded values within 10 kt).

Year	STS	NTSNZ	Industrial explosion and Azgir MEs	Kapustin Yar, Totsk, Aralsk regions	Total
1949	20				20
1950					0
1951	80				80
1952					0
1953	440				440
1954	80				120
1955	1,870	<10		40	1,880
1956	1,970			<10	1,980
1957	1,680	4,540		10	6,230
1958	80	16,130		20	16,230
1959					0
1960					0
1961	140	86,240		50	86,430
1962	220	132,710		900	133,830
1963					0
1964	90	20			110
1965	250				260
1966	420	1,400	30		1,850
1967	220	260	<10		490
1968	120	330	60		510
1969	270	540	60		870
1970	150	2,200	160		2,510
1971	300	2,450	130		2,880
1972	450	1,130	40		1,620
1973	310	7,820	40		8,170
1974	150	3,430	30		3,610
1975	210	4,190	20		4,420
1976	300	140	80		520
1977	350	130	50		530
1978	620	240	270		1,130
1979	960	280	170		1,410
1980	600	130	40		770
1981	610	140	70		820
1982	470	80	90		640
1983	440	250	90		780
1984	1,130	110	80		1,320
1985	450	0	10		460
1986	0	0	0		0
1987	1,000	150	40		1,190
1988	670	220	20		910
1989	300	0	0		300
1990	0	70	0		70

neutrons interact with construction components of a tower, barge, concrete platform, or other structures used for mounting or suspending nuclear devices before the explosion. The only principal difference between the two categories is that in the first, a harder neutron spectrum applies, and, besides the (n, γ) reaction, the threshold reactions of (n, 2n), (n, α), etc., types are quite possible (and in fact occur). In the second category the (n, γ) reaction mainly applies.

Twenty-two radionuclides with half-lives from 5.24 years to 1.5×10^7 years may be included in the IR category. Having considered the list, we conclude that ^{60}Co (half-life 5.24 years), ^{152}Eu (13.5 years), and ^{154}Eu (8.59 years) can be included, as well as ^{63}Ni (100 years), ^{93m}Nb (16.1 years), ^{121m}Sn (55 years), ^{151}Sm (90 years), ^{150}Eu (36 years), ^{178m}Hf (31 years), ^{192m}Ir (240 years), and ^{193}Pt (60 years).

Whether one of the listed radionuclides is recorded in significant quantities depends on the construction elements used for manufacturing the concrete nuclear explosion device. Alloyed steels often include such components as nickel, titanium, cobalt, and various rare earth elements. That is why IR sets in samples of soil taken after explosion should differ from one another.

Let us consider europium isotopes in more detail. Despite statements in many published sources, europium isotopes (^{152}Eu, ^{154}Eu, and ^{155}Eu) are both fission fragments and neutron reaction products. According to calculations for a surface 1-kt NE with energy release completely due to the fission reaction, 4.87×10^{17} nuclei of ^{152}Eu, 1.72×10^{19} nuclei of ^{154}Eu, and 8.7×10^{21} nuclei of ^{155}Eu are formed as a result of nuclear fission, and 3.7×10^{19} nuclei of ^{152}Eu, 1.1×10^{19} nuclei of ^{154}Eu, and 7.6×10^{16} nuclei of ^{155}Eu are formed as a result of neutron capture reactions. This estimate shows that ^{155}Eu can be considered mainly an FP radionuclide and ^{152}Eu can be considered an IR. As to ^{154}Eu, its content depends on explosion type and conditions of its conduct (one should remember that the estimate given above is for a device based only on fission reaction and detonated on the ground surface).

The third group, the TRs, is the largest contributor to environmental contamination. When ^{239}Pu is used as a nuclear explosive in the course of the explosion, the following reaction occurs: ^{239}Pu (n, γ), ^{240}Pu (n, γ), and ^{239}Pu (n, 2n). The relative effectiveness of the three reactions essentially depends on the exploded device design and the share of thermonuclear reactions in the total energy release. All of the mentioned plutonium

Table 9.7. Ratio of overall radioactivity to that of ^{152}Eu in large particles at the nearby trace of the thermonuclear explosion of 12 August 1953 (as of July 1995).

Specimen No.	^{90}Sr	^{137}Cs	^{154}Eu	^{155}Eu	^{60}Co	^{241}Am	^{238}Pu	$^{239+240}$Pu
0–53	–	16.00	0.040	0.440	0.120	3.17	2.67	92.7
1–53	–	7.02	0.077	0.232	0.190	3.17	2.87	31.3
2–53	–	15.42	0.060	0.350	0.140	2.20	5.86	62.3
10–53	29.5	6.51	0.100	0.359	0.200	3.35	–	47.4
12–53	19.3	5.24	0.171	0.391	0.140	2.87	2.14	25.7
13–53	–	5.35	0.091	0.377	0.065	3.62	4.41	29.9
Mean value	24.4		0.090	0.358	0.142	3.06	3.59	48.2

isotopes are almost always found in various ratios. These ratios are useful for separating the traces in a case of the superposition of the traces of different explosions. In the time that has passed since the last atmospheric test, the share of ^{241}Am (433 years) formed as a result of ^{241}Pu (14.35 years) beta-decay became predominant as compared to the cumulative input of the four plutonium isotopes.

Thus, it is obvious that isotopes of ^{238}Pu, ^{239}Pu, ^{240}Pu, and ^{241}Pu and ^{242}Am are the most significant and quantitatively detectable transuranic radionuclides. Attention should also be given to ^{242,243}Pu and ^{243}Am.

Both theoretical and experimental radionuclide composition estimates of the long-lived explosion products are presented in the report. Particles deposited 105 km from the craters of explosions on 29 August 1949 (the first nuclear test in the Soviet Union) and on 12 August 1953 (the first thermonuclear test in the Soviet Union) were analyzed. The data on radionuclide composition of the both explosions are presented in *Tables 9.7* and *9.8*.

As seen in *Table 9.8*, the relative yields of radioactive nuclides greatly differ in pure fission and thermonuclear explosions.

9.3.2 Radionuclide contamination of the STS and adjoining regions

According to requirements formulated in 1948, a nuclear weapons test site should be located in a desert area 200 km in diameter, adjacent to a railway station, and an airfield. A site 160 km from Semipalatinsk, bordered by the Shagan River (a tributary of Irtysh) and the Degelen and Kapyastan

Table 9.8. Ratio of overall radioactivity to that of ^{152}Eu in large particles at the nearby trace of the NE of 29 August 1949 (as of July 1995).

Specimen No.	^{90}Sr	^{137}Cs	^{154}Eu	^{155}Eu	^{60}Co	^{241}Am	^{238}Pu	$^{239+240}$Pu
2–49	–	12.61	0.027	0.030	0.294	0.139	0.038	15.50
3–49	33.5	2.61	0.031	0.100	0.210	0.050	0.234	1.64
9–49	–	1.20	0.028	0.037	0.057	0.018	–	1.94
14–49	–	8.66	0.056	0.264	0.468	0.223	0.300	32.00
15–49	157.4	25.26	0.012	0.485	0.640	0.346	0.400	47.00
Mean value	95.5	10.07	0.031	0.183	0.334	0.155	0.243	19.62

Mountains (100 km away from one another) met these requirements. The test site's initial area was about 5,200 km^2; its coordinates range from 49.7–60.125° north and from 77.7–79.1° east. Later its boundaries were widened. The schematic map of the test site today is presented in *Color Plate 4*.

The geographical position together with the primarily eastern (to the east, southeast, and northeast) direction of aerial mass motion (in frames of the general atmospheric circulation) predetermined the following regions as the most probable contamination areas at the Soviet territory (now Russian Federation and Republic of Kazakhstan): the Altay land, the Republic of Altay (Russian Federation), and the Semipalatinsk, East Kazakhstan, and Karaganda regions (Republic of Kazakhstan).

Numerous measurements of gamma-radiation at STS and adjoining areas, carried out by many institutions from 1991 to 1993 and compared with formerly conducted aero-gamma-radiometric and aero-gamma-spectrometric surveys (1963–1975), show that the probability was realized. The most significant traces, mainly from surface and low-altitude explosions, were recorded (outside the STS territory) just at the areas of the above-mentioned regions.

An aero-gamma-spectrometric survey carried out in 1991 by the State Scientific and Industrial Enterprise Aerogeologiya at STS and adjoining territories showed the presence of recorded radioactive traces with ^{137}Cs levels beyond $(1.11–1.85)\times10^{10}$ Bq/km^2 from NEs on 29 August 1949 (to 12–18 km along the axis from the epicenter of the explosion), 24 September 1951 (to 55–65 km), 12 August 1953 (to 80–95 km), and 15 January 1965 (to 13–18 km). Beginning with these distances, the traces are not contoured by a single isoline and are indistinguishable against the background of regional and global contamination.

From 1949 until 1962, 30 surface nuclear tests (namely, when the explosion fireball is in contact with the earth's surface) were carried out at STS. This total includes five cases when the nuclear device failed to work. As a consequence of such contact, huge amounts of soil are drawn into the explosion cloud and activated in it, and a radiometrically recorded trace of radioactive fallout is formed in the wind's predominant direction. For days, weeks, and even months, it is not difficult to observe such a trace by means of simple radiometric instruments at distances of hundreds of kilometers from the explosion's epicenter.

In a TsNIIAtominform official report, an estimate is given of the significance and potential effect of each of the 25 realized surface explosions and some air explosions conducted at the STS on all the territories adjacent to the test site. The work reports that the nuclear charge failed in five of the 30 surface nuclear tests conducted at the STS, and in 11 cases, energy release was less than 0.5 kt; the significant radioactive trace did not go out of the limits of the test site. The data on the tests are presented in the *Table 9.9*. The table also includes data on eight air explosions and an underground cratering explosion.

To evaluate relative significance of the traces on the basis of available primary data, the gamma-radiation exposure field was integrated in the trace part, limited by 0.1 R isoline for each surface test and the underground cratering explosion. It may be concluded from that evaluation that the surface explosions of 29 August 1949, 24 September 1951, 12 August 1953, and 24 August 1956 might qualify as NEs causing "very high" radioactive contamination (according to the authors' terminology). The surface explosions of 5 October 1954, 21 September 1955, 16 March 1956, and 7 August 1962 are to be attributed to the "high" contamination category. And the surface explosions of 29 July 1955, 2 August 1955, 5 August 1955, 25 March 1956, and 25 September 1962 are to be attributed to the "low" contamination category. In addition to air explosions with rather modest yields, four powerful air explosions (17 November 1956, 10 April 1957, 16 April 1957, and 22 August 1957) are included in *Table 9.9* because the tests were capable of producing high radioactive contamination in the remote zone. At the same time, trajectories of the spreading of air masses, carrying radioactive products of the explosions in the atmosphere, are not yet well known. In *Table 9.9*, data on the

Table 9.9. Dates and characteristics of some nuclear tests conducted at Semipalatinsk test site.

Date	Ground	Purpose	Energy release, kt	Height of burst, m	Main long-lived radionuclides activity, as of July 1994, Ci		
					^{90}Sr	^{137}Cs	Pu
Surface explosions							
29.08.49	P-1	NWR	22.0	30	500	1,400	360
24.09.51	P-1	NWR	38.0	30	900	2,550	300
12.08.53	P-1	NWR	400.0	30	22,000	29,000	280
05.10.54	P-3	NWR	4.0	0	100	280	105
19.10.54[a]	P-2	NWR	0	15	0	0	215
30.10.54	P-3	NWR	10.0	50	250	700	100
29.07.55	–	NWR	1.3	2.5	40	100	245
02.08.55	P-2	NWR	11.5	2.5	350	850	200
05.08.55	P-2	NWR	1.2	1.5	35	90	215
21.09.55	P-2	NWR	1.2	1.5	35	90	215
16.03.56	–	NWR	13.2	0.4	650	1,000	240
25.03.56	–	NWR	5.5	1	145	420	190
24.08.56	P-5	NWR	26.5	100	900	2,000	90
09.09.61	P-7	SAM	0.4	0	17	34	225
14.09.61	P-7	NWR	0.4	0	17	34	250
18.09.61	P-7	SAM	0.004	1	–	–	250
19.09.61	P-7	SAM	0.003	0	–	–	250
03.11.61[a]	P-7	SAM	0	0	–	–	230
04.11.61	P-7	NWR	0.15	0	4.5	13	195
07.08.62	P-5	NWR	10.0	–	370	870	200
22.09.62	P-3	SAM	0.2	0	7	17	280
25.09.62	P-5	NWR	7.0	0	270	610	205
05.11.62	P-1	WIE	0.4	15	16	35	190
11.11.62	P-3	NWR	0.1	8	3.2	8.6	210
13.11.62[a]	P-7	NWR	0	0	–	–	210
24.11.62[a]	P-7	SAM	0	0	–	–	140
26.11.62	P-7	SAM	0.03	0	–	–	210
23.12.62[a]	P-7	SAM	0	0	–	–	210
24.12.62	P-7	SAM	0.007	0	–	–	250
24.12.62	P-7	SAM	0.03	0	–	–	295
Air explosions							
18.10.51			~41.5				
23.10.54			~62				
22.11.55			~1,600				
17.11.56			150–1,500				
10.04.57			150–1,500				
16.04.57			150–1,500				
22.08.57			150–1,500				
17.01.58			1–20				
Underground cratering explosion							
15.01.65	Shaft 1004		140	Depth 178			

[a]Failure of nuclear charge.

Table 9.10. Tentative data on external exposure doses of the population (until the complete decay of radionuclides) in the area influenced by nuclear tests at the STS.

Region	Distance from the site, 1,000 km	Pop., 1,000 persons	Max. external exposure dose, cSv	Mean external dose of pop., cSv	Collective external exposure dose, 1,000 person-Sv
Altay land	0.14–0.7	2,514	52.00	0.50	13.50
Republic of Altay	0.40–0.8	174	0.50	0.20	0.30
Republic of Khakassia	0.70–1.0	508	0.20	0.15	0.76
Novosibirsk region	0.50–0.7	2,657	1.00	0.05	1.44
Kemerovo region	0.70–1.0	2,990	1.00	0.06	1.64
Krasnoyarsk land	0.90–2.2	600	0.12	0.04	0.24
Irkutsk region	1.30–2.7	1,340	0.10	0.04	0.47
Chita region	2.00–3.0	1,258	0.05	0.04	0.44
Tomsk region	0.70–1.3	887	0.15	0.04	0.35
Total		15,928			19.14

underground cratering NE, conducted on 15 January 1965 in Shaft 1004 are presented. The explosion was carried out to form a large crater, and subsequently filled with water during the spring flood. The height of the upper edge of the formed cloud was 4,800 m. In the course of wind drift of the cloud, it was "torn." Its lower part, containing the main part of radionuclides, moved across the Altay land territory, and Barnaul city was caught in its train. Estimates of Dubasov *et al.* (1996a) show that the explosion may be assigned to the high contamination category.

The main biologically significant nuclides in the territory contaminated by radioactive fallout as a result of the air nuclear tests are ^{90}Sr, ^{137}Cs, and Pu. The total activity of the radionuclides after the tests at the STS (taking into account the radionuclide decay), as of 1 January 1996, is as follows:

- ^{137}Cs (gamma-activity): $\sim 1.48 \times 10^{15}$ Bq.
- ^{90}Sr (beta-activity): $\sim 10^{15}$ Bq.
- Plutonium (alpha-activity): $\sim 1.45 \times 10^{14}$ Bq.

An estimate of the exposure dose from nuclear testing at STS on the population of nine regions of Russia are presented in *Table 9.10*.

9.3.3 Radionuclide contamination of the NTSNZ and adjoining regions

Nuclear tests at the NTSNZ were conducted at three technological grounds which we will term zone A, B, and C (*Figure 9.1*).

Zone A (Area of Chernaya and Bashmachnaya Inlets)

Three underwater and two above-water tests (1955–1962), a surface NE of 7 September 1957, and six underground NEs in shafts (1972–1975) were conducted at the ground. In *Figure 9.1* the radioactive contamination spots are marked with numerals and defined below:

1. The trace of the 1957 NE. In the crater formed at the coast of Chernaya inlet, the gamma-radiation dose rate does not exceed 400 μR/hr. In the rest of the trace area (well 10 m from the crater), it is 80 μR/hr, and sharply decreases with distance from the crater's center. The crater's location, together with the adjacent contaminated lands, has the status of a sanitary-protection zone (SPZ). As of 1 January 1996, its area did not exceed 0.5 km^2.
2. A spot formed as a result of 1955 underwater NE at Kushny Peninsula. Now the size of area is less than 0.2 km^2, and the gamma-radiation dose rate is less than 25 μR/hr. The trace stretches to the south from the bay into the sea.
3. The location of the 1961–1962 air NE test series, 14 km northwest of the Potych Cape. The size of the area is less than 0.2 km^2, and the gamma-radiation dose rate is less than 25 μR/hr.
4. The trace of a former outlet of steam and gas contaminated with radioactive products from a 1973 underground NE. The area is now less than 0.3 km^2, and the gamma-radiation dose rate is less than 20 μR/hr. The trace is narrow and spotty.
5. A trace in the form of a broken spot resulting from a 1961 above-water NE, stretches from the bay to the northeast. The area is less than 0.4 km^2, and the gamma-radiation dose rate is less than 25 μR/hr.

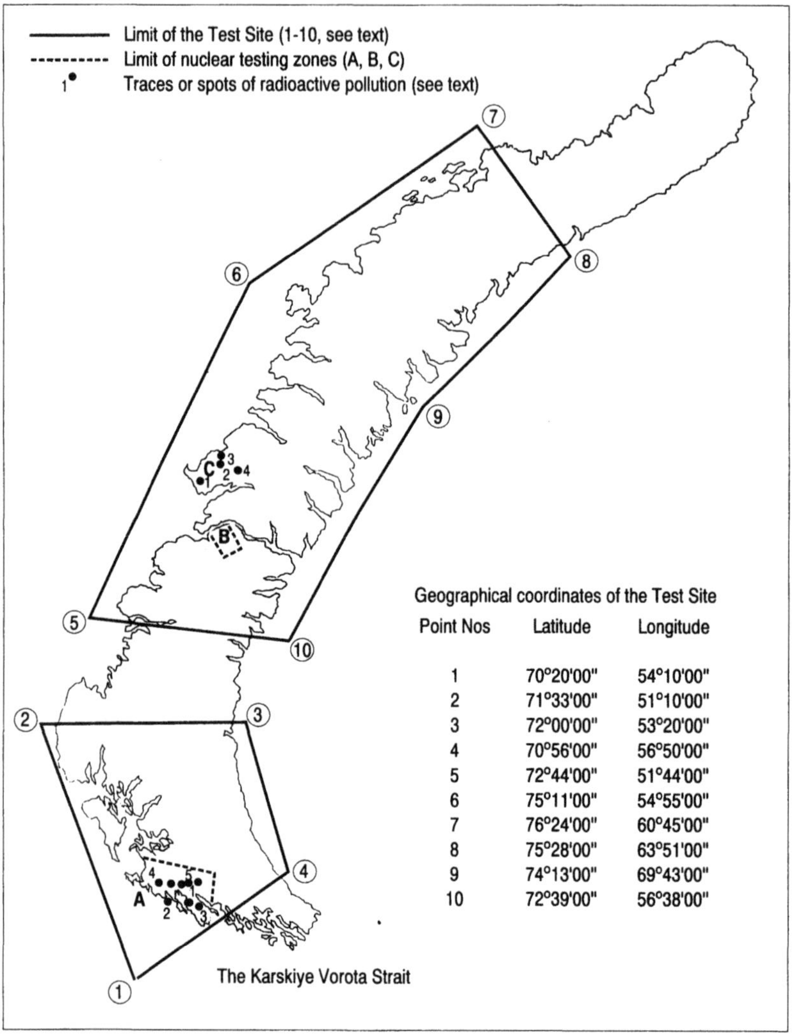

Geographical coordinates of the Test Site

Point Nos	Latitude	Longitude
1	70°20'00"	54°10'00"
2	71°33'00"	51°10'00"
3	72°00'00"	53°20'00"
4	70°56'00"	56°50'00"
5	72°44'00"	51°44'00"
6	75°11'00"	54°55'00"
7	76°24'00"	60°45'00"
8	75°28'00"	63°51'00"
9	74°13'00"	69°43'00"
10	72°39'00"	56°38'00"

Figure 9.1. Novaya Zemlya northern test site.

The natural background level, outside the places noted for their higher radiation dose rate, is 7–11 μR/hr.

Zone B (Area in the Western Part of the Matochkin Shar Strait)

Thirty-six NEs in tunnels were conducted at these grounds. Some places have radioactive noble gases that flowed directly to the earth's surface after underground NE detonations. There are local plots with higher gamma-radiation levels that are 1.5–2.0 times above the background level. There is a separate spot (tunnel A-37A), less than 0.3 km^2 in size, with a maximum gamma-radiation level of about 500 μR/hr, just at the point of a former outlet of radioactive gases. A SPZ is being preserved in the A-37A tunnel area. In the rest of zone B territory, as well as at the settlement, the gamma-radiation dose rate is 7–12 μR/hr.

In the course of conducting underground NEs in zone B, two abnormal radiation situations took place. The first case happened on 14 October 1969 in A-9 tunnel. About 60 minutes after detonation, a sudden break occurred along a tectonic crack in the epicenter zone and a steam and gas mixture was released. At the technological ground, the dose rate reached several hundred R/hr. On the third day, a slow transfer of radioactive products began from the test site territory to the north-northwest into the Barents Sea, where their detection was possible 500 km away. There was no radioactive fallout.

The second abnormal situation took place on 2 August 1987 in A-37 tunnel. About 1.5 minutes after detonation, a break occurred along a natural rupture of an undermelted glacier at the mountain's slope along the tunnel's axis and a steam and gas mixture was released. Besides the radionuclides of noble gases, radionuclides of barium, iodine, cesium, strontium, antimony, tellurium, and others drifted into the atmosphere. For six days the radioactive products were in the technological ground area. As a result, the dose rate at the checkpoint exceeded 500 R/hr. There was no radioactive fallout outside the test site, with the exception of trace amounts of radioiodine.

Zone C (Area of Sukhoy Nos Peninsula)

Until 1962, NEs were conducted in four locations in this zone:

1. The western spot, 3 km east of Fedorov Mountain (12 km from Gagachiy Island). The area is about 0.5 km^2; the exposure rate exceeds the background level by two times.
2. The central spot, located in the peninsula's center (1.5 km north of two lakes). The area is about 0.3 km^2; the exposure rate exceeds the background level by two times.

3. The northern spot, located 10 km from Tsivolki Cape, along the azimuth of 150°. The area is about 0.3 km^2; the exposure rate exceeds the background level by two times.
4. The eastern spot, located 12 km northeast of the Klochkovskii Peninsula (between hills 257 and 249). The area is about 0.4 km^2; the exposure rate exceeds the background level by two times.

The contamination of the test site territory (excluding SPZs) by ^{137}Cs is 0.04–0.18 Ci/km^2 (0.9 Ci/km^2 on average), and that contaminated by ^{90}Sr is 0.04–0.06 Ci/km^2; the exposure rate is 7–12 μR/hr. These values hardly differ from contamination density and radiation background level in the middle latitudes of the Northern Hemisphere. The highest contamination is observed in a limited area (about 1 km^2) around a surface NE (1957) crater on the Chernaya inlet seashore. There the values of ^{137}Cs contamination density vary from 0.01 Ci/km^2 to 10 Ci/km^2.

In the Chernaya inlet, three underwater and two above-water NEs were conducted. In 1993, data became available from joint studies of specialists of the Murmansk Institute for Biology of Northern Seas and Lublin University (Poland) that was devoted to analysis of samples of bottom sediments from the Chernaya inlet. According to the data of gamma-spectrometric analysis, ^{137}Cs activity in the bottom sediments is 1,444.2 Bq/kg, and ^{242}Am activity is 2,662.0 Bq/kg.

An estimate of the exposure doses from nuclear testing at NTSNZ on the population of 21 regions of Russia is presented in *Table 9.11*.

9.3.4 Nuclear explosions at the missile testing range

The missiles with nuclear and thermonuclear warheads to be tested in outer space were launched from the missile testing range (MTR) located near Kapustin Yar, Astrakhan region, Russian Federation. All the explosions that took place from the MTR are listed in *Table 9.12*. None of the NEs in *Table 9.12* contributed any contamination to the MTR territory and adjacent regions, as all of the detonations were conducted at high altitudes.

Besides the explosions mentioned in *Table 9.12*, one additional test should be noted: A missile with a nuclear warhead was launched from the MTR on 2 February 1956 and the charge detonated near Aralsk, Kazakhstan. Information concerning radioactive contamination of that site is still classified and unavailable.

Table 9.11. Tentative data on external exposure dose of the population (until the complete decay of radionuclides) of various regions of Russia in the area influenced by nuclear tests at the NTSNZ.

Region	Distance from the site, 1,000 km	Pop., 1,000 persons	Max. external exposure dose, cSv	Mean external dose of pop., cSv	Collective external exposure dose, 1,000 person-Sv
Krasnoyarsk land (ADs not included)	1.3–3.0	2,693	0.7	0.10	3.0
Taimyr (Dolgano-Nenets) AD	0.9–2.2	48	2	1.0	0.5
Evenk AD	1.6–2.4	17	1.5	0.7	0.12
Republic of Sakha (Yakutia)	2.0–3.7	883	1	0.8	7.0
Tyumen region (ADs not included)	1.8–2.2	1,165	0.3	0.15	1.8
Yamalo-Nenets AD	0.5–1.8	193	0.4	0.13	0.25
Khanty-Mansi AD	0.9–1.9	673	0.3	0.17	4.9
Perm region (ADs not included)	1.3–2.0	2,830	0.3	0.17	4.9
Magadan region (together with the Chukotsk AD)	3.8–4.5	490	0.6	0.25	1.2
Republic of Komi	0.8–1.6	1,147	0.4	0.17	2.0
Khabarovsk land	3.6–4.5	1,610	0.6	0.2	3.2
Nenets AD of Arkhangelsk region	0.4–0.8	50	0.3	0.10	0.5
Republic of Udmurt	1.7–2.0	1,516	0.2	0.11	1.6
Sverdlovsk region	1.4–2.0	4,500	0.3	0.20	9.5
Kurgan region	2.0–2.2	1,085	0.2	0.14	1.5
Chelyabinsk region	2.0–2.4	3,480	0.2	0.14	4.8
Republic of Bashkortostan	2.0–2.4	3,865	0.2	0.10	4.0
Omsk region	1.9–2.4	1,963	0.15	0.10	2.0
Republic of Tatarstan	1.9–2.2	3,453	0.15	0.06	2.4
Irkutsk region	2.6–3.4	2,616	0.3	0.005	0.8
Chita region	3.4–3.9	1,258	0.2	0.001	0.15
Total	–	35,535	–	0.15	52.27

AD = Autonomous district.

Table 9.12. NEs, conducted in the atmosphere and in space using missiles launched from the MTR.

Number in the NE list	Date	Type	Purpose	Energy release, kT
34	19.01.57	Air	WIE	10
82	01.11.58	Air	WIE	10
83	03.11.58	Air	WIE	10
88	06.09.61	Air	WIE	11
115	06.10.61	Air	WIE	40
127	27.10.61	Space	WIE	1.2
128	27.10.61	Space	WIE	1.2
184	22.10.62	Space	WIE	300
187	28.10.62	Space	WIE	300
195	01.11.62	High altitude	WIE	300

9.3.5 1954 military exercises in Totsk

The explosion of a nuclear bomb with an energy release of 40 kt was conducted at a height of 350 m, and the scaled HOB was 102 m/kt$^{1/3}$. The fireball did not touch the underlying surface, and fission products as well as residual plutonium dropped out on the remote trace. As a result of absorption of neutrons by the soil layer, induced activity of various radionuclides (including ^{60}Co, ^{152}Eu, and ^{154}Eu) formed in the epicenter of the explosion. A dust column raised from the epicenter zone, containing the induced radionuclides, dropped out on the near trace to an extent of 210 km. The maximum cumulative dose, about 1 rem, has been absorbed on the trace up to 70 km from the epicenter.

On 28–29 July 1994, in preparation for joint exercises of peacekeeping forces at the Totsk Firing Ground, a radiological examination was carried out. The examination was conducted by a team of specialists in the fields of radiation safety, radiation protection, and radiobiology, including four Russian representatives and four American representatives.

At the first stage of the work, 80 measurements of gamma-radiation dose rates were carried out, and 38 samples of soil were taken by the Russian specialists to conduct further investigations in laboratories. The samples were taken at different points, including those along azimuth 79° at 200 m, 300 m, and 650 m from the epicenter and along azimuth 259° at 30 m and 100 m from the epicenter. The results of gamma-spectrometric analysis of a sample (azimuth 259°, 100 m from the epicenter) are as

Table 9.13. Preliminary data on external exposure dose of the population (until the complete decay of radionuclides) in the area, influenced by the NE of the 1954 Totsk military exercises

Region	Distance from the firing ground, 1,000 km	Pop. in the contaminated territory, 1,000 persons	Max. external exposure dose, cSv	Mean external exposure dose of the pop., cSV	Collective external exposure dose, 1,000 person-Sv
Near zone (Orenburg region)	0–0.2	20	1	0.3	0.065
Remote zone (Krasnoyarsk land)	1.5–2.1	150	0.12	0.05	0.081
Total area	–	170	–	0.175	0.146

follows: ^{152}Eu, 217 Bq/kg; ^{154}Eu, 8.9 Bk/kg; ^{60}Co, 8.5 Bq/kg; ^{137}Cs, 58 Bq/kg; ^{134}Cs, 1.5 Bq/kg; and ^{40}K, 338 Bq/kg. The first three radionuclides are the Totsk 1954 explosion products (^{134}Cs is of Chernobyl origin; ^{137}Cs originated in global fallout and partly from Chernobyl).

An estimate of the Totsk explosion effect on the population of adjacent regions of Russia is presented in *Table 9.13*.

9.3.6 Peaceful NEs in the Soviet Union

The Soviet Union began an extensive program of underground NEs for use in peaceful activities in 1965. A total of 124 PNEs (135 exploded nuclear charges) were conducted and 32 tests (38 exploded nuclear devices) were carried out with PNE technologies. The program started with a cratering explosion in shaft No. 1004 at STS on 15 January 1965, and it was stopped after the NE Rubin, held on 6 September 1988 in the RN-1 shaft in Arkhangelsk region of RF.

The program of PNEs included the following main activities:

- Deep seismic probing of the earth's crust to search for promising mineral resources.
- Experimental work on oil extraction intensification and improved oil recovery.
- Experimental work on gas extraction intensification.
- Experimental studies of technology for cavity production in rock salt.

Table 9.14. Soviet underground NEs conducted in the interests of the national economy from 1965 to 1988, by energy release.

Energy release, 1,000 t TNT	Technological ground name (number of PNE)	Total
Less than 1	Klivazh, Lazurit, Tavda, Telkem (2)	5
1–3	Benzol, Butan, Vega, Geliy, Globus, Dniepr, Kristall, Neva, Region (2), Sovkhoznoye, Sary–Uzen, Galit	13
3–5	Dniepr, Fakel	2
5–10	Agat, Batolit (2), Butan, Vega (13), Vyatka, Galit, Gorizont (4), Grifon (2), Kama (2), Kimberlit (2), Meridian (3), Meteorit (3), Region (3), Rift (2), Rubin, Takhta–Kugultinskoye, Shpat	43
10–20	Angara, Galit, Dedurovka (2), Kvarts (3), Kimberlit, Krater, Kraton (4), Lura (6), Meteorit, Neva (3), Oka, Rift, Rubin, Sheksna	27
20–50	Galit (2), Pamuk, Pirit, Say–Utyos, Taiga, Urta Burlak	7
50–100	Galit (5), Say–Utyos (2)	7
100–150	Chagan	1

- Experimental work on creating underground reservoirs.
- Testing of PNE technology.
- Experimental work on gas spouter shaft closure.
- Experimental work on creating trench pits in alluvial soils.
- Experimental work on dam creation by loosening of rocks.
- Experimental work on burial of biologically hazardous petrochemical industrial wastewaters in deep geological structures.
- Experimental work on preventing sudden coal dust and methane outbursts.
- Testing of ore fragmentation technology.

Data on PNE locations, energy releases, depths of charge burial, and number of exploded charges are presented in *Tables 9.14* and *9.15*.

While carrying out the PNE program, five cratering explosions were conducted, four of them at the STS and one (Taiga) at the border of Perm region and Komi Republic, 6 km north of Chusovskoye Lake. That experimental NE salvo was conducted on 23 March 1971 to study the possibility of constructing the Pechora-Kolva canal as a part of the project to transfer

Table 9.15. Soviet underground NEs conducted in the interests of the national economy from 1965 to 1988, by depth of charge burial.

Depth of charge burial, m	Technological ground name (number of PNE)	Total
Less than 300	Galit, Dniepr (2), Kristall, Lazurit, Sary-Uzen, Tavda, Taiga, Telkem (2), Chagan	11
300–500	Globus, Meridian (2), Region (4), Sary-Utyos	8
500–800	Agat, Batolit, Galit, Globus (3), Gorizont (3), Kvarts (3), Kimberlit, Kraton (3), Meridian, Meteorit (3), Region, Rift, Rubin, Sary-Utyos (2), Sovkhoznoye, Takhta-Kugultinskoye, Shpat	28
800–1000	Batolit, Vega (5), Galit (7), Gorizont, Klivazh, Kimberlit (2), Kraton, Lira (6), Meteorit, Neva, Rift (2), Rubin	29
1,000–1,500	Butan (5), Vega (10), Galit, Grifon (2), Dedurovka (2), Pirit	21
1,500–2,000	Vyatka, Krater, Neva (3), Oka, Urta-Bulak, Sheksna	8
2,000–2,500	Angara, Benzol, Galit (5), Kama (2), Pamuk, Fakel	11

water from northern rivers to the drainage basin of the Volga River. Some 200–300 km from the crater's edge, the gamma-radiation exposure rate begins to rise in the direction toward the epicenter. Maximum values of measured exposure rate were 2,500 and 840 μR/hr. The points are located in close vicinity to the crater and within zones where the exposure rate level is higher than 500 μR/hr.

In addition to the above-mentioned cratering explosions, where the release of some activity into the atmosphere was predicted, in some cases, several emergency releases of radioactive products to the earth's surface have also occurred.

In particular, as a result of NE Kristall (2 October 1974) and Kraton-3 (24 August 1975) in Yakutia, part of the radionuclides released into the atmosphere caused radioactive contamination of some lands with the spread of the radioactive substances. Some release of gaseous products into the atmosphere was expected at that object, in accordance with the project's goal of creating an earth-fill dam through the use of a burial charge. The process of the NE gaseous products filtration through the loosened rocks was controlled, and it stayed within the parameters of the predictive models. Radiation safety of the work was ensured by corresponding technical measures, choice of favorable meteorological conditions, and continuous radiation monitoring. At Kraton-3, an unforeseen release of radionuclides into the atmosphere occurred due to improper grouting of holes. In that case, the release of radioactive gases was recorded at a distance up to 150 km in the uninhabited forest and tundra area.

A short description of the radiation situation at these objects is presented below.

Object Kristall. The epicenter zone includes a hill in the form of a truncated cone about 160 m in diameter and up to 8 m in height. It is now covered with tundra vegetation. According to data of airborne gamma survey, made by the Central Search and Survey Expedition of Production and Geological Association Yakutskgeologiya in 1990, at a 0.4×0.9-km plot near the underground NE epicenter, gamma-radiation exposure rates mainly ranged from 15–30 μR/hr; the maximum value was 110 μR/hr. In the cover soil, ^{60}Co and ^{137}Ce were detected and identified in trace amounts (less than 50 Bq/kg). In water samples taken from the pile, no radionuclides were detected (the detection methods' sensitivity being less than 0.1 Bq/l for ^{137}Ce, and less than 100 Bq/l for ^{3}H). As the site is in a strongly propagated permafrost area, the migration of radionuclides with underground waters is very complicated.

The site is practically nonhazardous to the population and the environment. Still, complications may arise if uncontrolled digging or drilling into the pile occurs. Therefore, earthwork and drilling on the pile and within the range of 100 m are prohibited. In addition, some sections of the pile were filled with a 1.5-m-thick clean soil layer that led to a decrease of the radiation level on the surface. Covering of the pile with clean soil also prevents the direct contact of radionuclides with people. It was also recommended that the SPZ regime be kept provisionally in the pile area with periodic monitoring of environmental objects.

Object Kraton-3. According to measurement data gathered in 1990, a radioactive trace was recorded up to 5 km from the NE epicenter. Its width was from 0.5–2.5 km; the maximum level of gamma-radiation was up to 200 μR/hr along its axis and up to 730 μR/hr in the epicenter (the shaft collar). At present, at the limits of the remediated area, the gamma-radiation exposure rate is 30–50 μR/hr on average. Contamination of the soil-vegetation cover with beta-emitting radionuclides along the 2-km-long trace is an additional radiation factor at the Kraton-3 site. It was recommended to keep the SPZ regime in the area with periodic radiation monitoring and restriction of use.

Abnormal radiation situations took place during three other PNEs: Globus-1 (19 September 1971) in Ivanovo region, 40 km east northeast of Kineshma; and A-1 (22 April 1966) and A-8 (17 January 1979) in Kazakhstan, 180 km north of Astrakhan. In these cases the exposure rate outside the technological grounds was at the global background level, and in the work shaft zone it was from 15–60 μR/hr.

9.3.7 Global radioactive fallout

^{90}Sr and ^{137}Cs, as well as ^{3}H (to a degree), are the most significant isotopes in global fallout due to their long life and, consequently, potential hazard. The observed distribution of global contamination levels of ^{90}Sr and ^{137}Cs is determined by a number of factors. The most essential of them are as follows:

- Nuclear weapons test characteristics (energy release, type, purpose, etc.).
- The number of NEs, mainly in the Northern Hemisphere.
- General regularities of atmospheric circulation.
- Meteorological and climatic peculiarities of a region.

According to data of the UN Scientific Committee on Atomic Radiation Effects (UNSCEAR), from the first nuclear weapon tests until 1963, 0.71×10^{18} Bq of ^{137}Cs was formed and released into the atmosphere. The contribution from later tests was less significant. If this amount of radionuclides were distributed uniformly over the earth's surface, the following levels of contamination would have been observed in 1970: 30 mCi/km^2 for ^{90}Sr and 53 mCi/km^2 for ^{137}Cs. The actual contamination levels result from the action of all the above-mentioned factors, and, in some places, they differ greatly from the mean values.

^{90}Sr distribution in soils was studied systematically. Until 1963, the amount of ^{90}Sr in soils grew rapidly. By 1964, a significant part of the airborne ^{90}Sr (about 3.7×10^{17} Bq) dropped onto the earth's surface. Later, after cessation of air NEs in the USA and in the Soviet Union, the rate of ^{90}Sr fallout was largely compensated by its radioactive decay and transport to seas and oceans with river flow. This may be illustrated by the following data on mean values of ^{90}Sr contamination of the territories of the former Soviet Union (FSU):

Years	1963	1964	1965	1966	1967	1968	1969
^{90}Sr content, mCi/km^2	44	56	54	53	44	43	43

The results were received from about 40 points located in different climatic zones of the country.

Detailed information concerning ^{137}Cs global fallout distribution in the 1960s is available only for those areas of the FSU where detailed gamma-surveys were conducted, using techniques described in official documents.

The distances between flight strips were 50 km, as a rule. In some cases they reached 100 km. The mean concentration of radionuclides along the flight line was usually determined for sections of about 10 km. The flight altitude was 25–50 m, so the resulting data specify a 100- to 200-m-wide strip on the ground (its area is about 1 km^2). The error rate of the reproducibility of ^{137}Cs content determination, taking into account uncertainty in the navigational accuracy, was not more than 10%. The error rate of ^{137}Cs content determination was about 20% at the contamination level of about 0.1 Ci/km^2.

Figures 9.2 and *9.3* present maps of territory in the FSU with global fallout contamination (by the late 1960s) from NEs. The contamination levels mainly range from 50–175 mCi/km^2; the mean ^{137}Cs content is 90 mCi/km^2.

Particular attention is paid to the extremely complicated pattern of contamination of the territory with ^{137}Cs and the presence of numerous local areas with contamination levels quite different from those of surrounding territories. A somewhat similar spotty pattern can be also observed in the map of fallout.

9.4 Existing Databases

Information is available from the Ministry of Defense on all nuclear tests conducted on Novaya Zemlya, including data on the types of NEs, radiation situation while conducting tests, radiation consequences of each test,

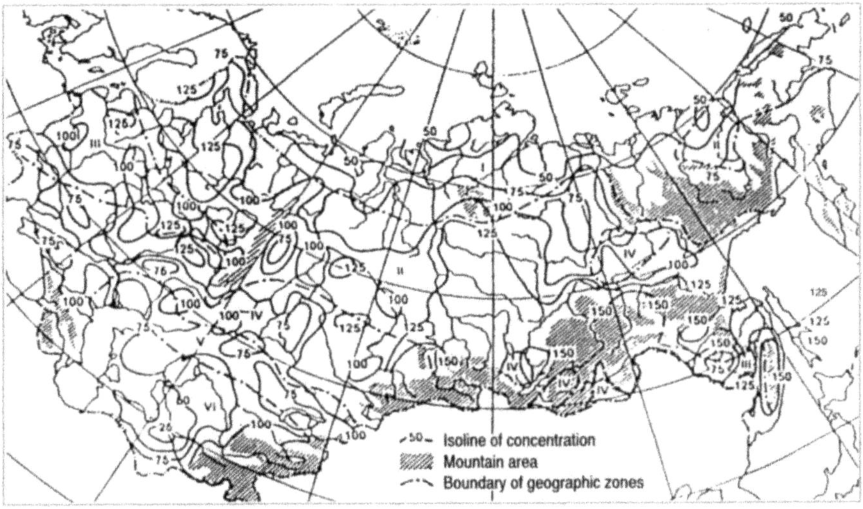

Figure 9.2. ^{137}Cs contamination over Soviet territory in the late 1960s, mCi/km^2: I–tundra and forest-tundra; II–coniferous forest; III–mixed forest; IV–forest steppe and steppe; V–semidesert; VI–desert.

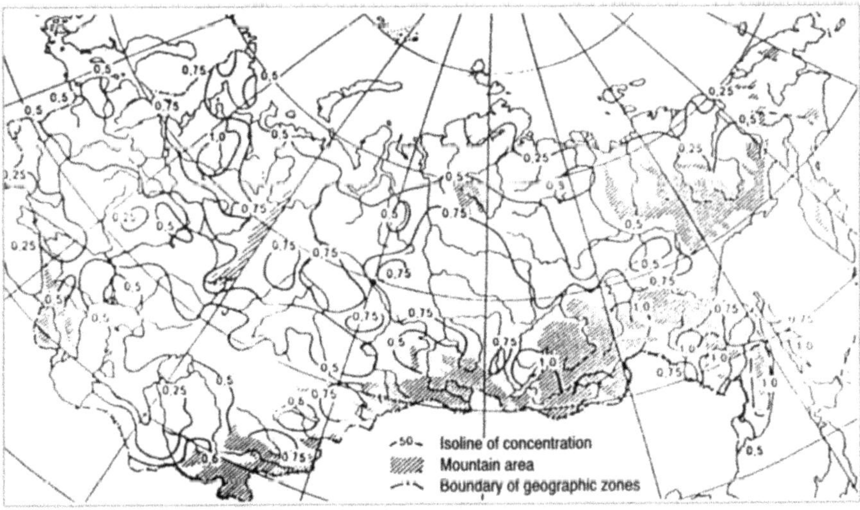

Figure 9.3. Contamination over Soviet territory in the 1970s: gamma-exposure rates (μR/hr) at a height of 1 m (1974 survey).

and the dynamics of the radioecological situation on Novaya Zemlya over 30 years.

Some of the data are still classified, including data on parameters of tested charges and devices, some data on testing conditions, and an insignificant part of data concerning the radiation situation after the test.

Some fields in the tables are blank concerning particular tests because, at that time, it was impossible to measure some elements by any technical means.

The following data are available:

- The chronology of tests.
- Estimates of the total amount of airborne radionuclides and amounts of separate radionuclides (calculated and measured data).
- Parameters of the radiation situation while conducting nuclear tests.
- The testing conditions.
- Exposures at the test site and along the route of the radioactive trace.
- Dynamics of some radionuclides accumulating in environmental objects at the test site (vegetation, fauna, water reservoirs, soil, etc.).
- Radiation situation before each test (background).
- Dynamics of the radioecological situation at the test site after testing, etc.

In addition to the data of the Ministry of Defense, some data are available from other departments that took part in conducting the tests. On the basis of instrumental studies, the following estimates have been made:

- Total amount of airborne radionuclides formed as a result of all tests.
- Basic exposures in the regions for all years.
- Contamination of environmental objects.
- Other estimates of radiation consequences.
- Estimates of cavities in mining works that might be used for RW storage or adapted for their disposal.
- Radioecological capacity of a test site.

Bibliography

Andryushin, I.A., Bogdan, V.V., Vaschinkin, S.A., Zelentsov, S.A., Zolotukhin, G.E., Karimov, V.M., Kirichenko, V.V., Matushchenko, A.M., Silkin,

Yu.A., Strukov, V.G., Kharitonov, K.V., Tchernyshev, A.K., Tsyrkov, G.A., and Shumaev, M.P., 1996, *USSR Nuclear Weapons Tests and Peaceful Nuclear Explosions: 1949 Through 1990*, RFNC–VNIIEF (Russian Federal Nuclear Center–the All-Russia Research Institute of Experimental Physics), Sarov, Russia, ISBN 5-85165-062-1.

Boltneva, L.I., Israel, Yu.A., Nazarov, I.M. *et al.*, 1977, ^{137}Cs and ^{90}Sr global contamination and external exposures at the territory of the USSR, *Atomic Energy*, **42**:355.

Dubasov, Yu.V., Krasilov, G.A., Matushchenko, A.M. *et al.*, 1996a, The chronology of air nuclear explosions at the Semipalatinsk Test Site and their radiation characterization, *Bulletin of the Public Information Center on Atomic Energy*, **6**:39–46, TsNIIAtominform, Moscow, Russia.

Dubasov, Yu.V., Matushchenko, A.M. *et al.*, 1996b, Underground explosions of nuclear devices at the territory of the USSR in 1965–1988: The chronology and radiation consequences, *Bulletin of the Public Information Center on Atomic Energy*, **6**:39–46, TsNIIAtominform, Moscow, Russia.

Dyachenko, V.I., Kazantsev, V.V., Martakov, Yu.P., and Semenovykh, S.V., 1995, Radiation situation in the area of nuclear explosion, conducted at the Totsk Firing Ground on September 14, 1954, *Bulletin of the Public Information Center on Atomic Energy*, **5–6**:44–47, TsNIIAtominform, Moscow, Russia.

Israel, Yu.A., 1996, *The Radioactive Fallout After Nuclear Explosions and Accidents*, Progress-Pogoda, St. Petersburg, Russia.

Israel, Yu.A., Stukin, E.D. *et al.*, 1994, Reconstruction of the actual pattern of ground radioactive contamination as a result of nuclear tests and accidents, *Meteorologiya and Gidrologiya*, **8**:5–18.

Kogan, R.M., Nazarov, I.M., and Fridman, Sh.D., 1969, *The Fundamentals of Gamma-spectrometry of Natural Environment*, Atomizdat, Moscow, Russia.

Kruglov, A.K., 1994, *How the Nuclear Industry was Created in the USSR*, TsNIIAtominform, Moscow, Russia.

Logachev, V.A., 1996, Estimation and comparison of radiation exposure of the population of Russian Federation as a result of conducting air nuclear explosions at the test sites of ex-USSR and of the Chernobyl accident, *Report presented to the International Conference "One Decade After Chernobyl: Summing up the Consequences of the Accident,"* 8–12 April, Vienna International Center, Vienna, Austria.

Morozov, A.A., Reshetov, V.V., and Terminasov, S.Yu., 1994, The results of inspection of nuclear explosion location, In: *Methods and Some Results of Airborne Gamma-survey of Radioactive Contamination of the European Part of Russia*, pp. 230–237, Gidrometeoizdat, St. Petersburg, Russia.

NTSNZ (Northern Test Site Novaya Zemlya), 1992, *Nuclear Explosions in the USSR*, Issue No. 1, NTSNZ, Moscow, Russia.

Russian Research Institute for Cultural and Natural Legacy – The Foundation of Polar Studies, 1996, *Radiological situation at the Central Test Site of the Russian Federation, In Novaya Zemlya: Natural and Cultural Legacy – The History of Discoveries*, pp. 156–157, Russian Research Institute for Cultural and Natural Legacy – The Foundation of Polar Studies, Moscow, Russia.

Stukin, E.D., 1995, *Feasibility Study of Retrospective Reconstruction of Traces of Nuclear Explosions, Conducted at the Semipalatinsk Test Site, Contributing to Contamination of Investigated Altay Land Territory*, IGKE (Institute of Global Climate and Ecology), Moscow, Russia.

TsNIIAtominform, 1994, Peaceful use of underground nuclear explosions, *Nuclear Explosions in the USSR*, Issue No. 4, TsNIIAtominform, Moscow, Russia.

Chapter 10

Storage and Reprocessing of Nonreactor Radioactive Wastes and Spent Ionizing Radiation Sources

10.1 Introduction

A nation's development of a modern, industrialized economy is inevitably connected to the formation of wastes, including those containing radioactive substances. Among the radioactive wastes (RWs) are materials not to be used later, including solutions, products, equipment, biological objects, and soils containing radionuclides exceeding safety standards.

Nonreactor RWs of low and intermediate specific activity are formed as a result of radionuclides used in medicine, industry, agriculture, and scientific research. In the former Soviet Union (FSU), the basic suppliers of such wastes include numerous research organizations, industrial enterprises, medical establishments, and educational institutions, which, as a rule, are located in densely populated regions. These enterprises send their RWs and spent ionizing radiation sources to 16 regional points of burial of RWs (PBRW) forming a system called Radon Special Enterprises (SEs), with each SE located near a large industrial center. A region of service is determined for each SE. The zones of service and the arrangement of the SEs are shown in *Figure 10.1*.

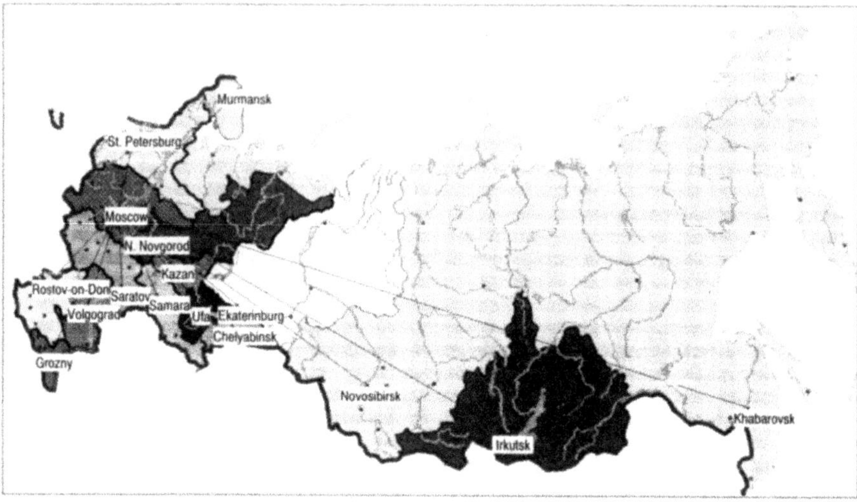

Figure 10.1. Location and zones of servicing for Radon SEs.

The Moscow Division of the Scientific Industrial Association (SIA) Radon is the principle division among the regional divisions as regards the handling of RWs of intermediate and low specific activity in the Radon system. Scientific, practical, methodical, and other help is provided by SEs of the Russian Federation (RF) and, if necessary, by the countries in the Commonwealth of Independent States.

Within the framework of the scientific-methodological activity of SIA Radon, the data collection and systematization on all kinds of SE activity, including radioecological monitoring of environmental objects, are carried out.

The list of registered parameters comprises:

- Title and location of object.
- Geographical and geomorphological characteristics of object location.
- Regions of service.
- Type and project of storehouses, and date of commissioning.
- Volume of storehouses and percentage of filling at any given time.
- Volume of annual receipt and total activity of the wastes since burial.

Technological processes used in handling the RWs include:

- Cementing, which is carried out on an industrial scale at stationary and mobile plants.
- Bituminization, in which RWs with concentration more than 60 g/l are vaporized five to six times and then incorporated into melted bitumen at a facility using a rotor vaporizer.
- Burning and pressing, which is carried out on industrial-type installations. Additionally, development of high-temperature methods of RW burning in pit ovens with fuel-plasma sources of heating is conducted now at SIA Radon.
- Vitrification, which is a new, highly reliable method of RW localization, produces an end product suitable for final disposal.
- Water cleaning is carried out by both stationary and mobile facilities.
- Solidification, in which spent ionizing radiation sources relating to a category of high-level RWs are dumped in a special depository at a depth of 6 m and, using a Moskit machine, are covered, layer by layer, with molten metal.

Except for the Moscow SIA Radon, which has an advanced infrastructure, the hardware and technological equipment at the majority of PBRWs do not correspond to modern requirements. Capacities of storehouses are exhausted or extremely limited. New sites for RW disposal are required for a number of PBRWs.

The general characteristics of the Radon SEs in Russia are given in *Table 10.1*.

10.2 Characteristics of Accumulated Radioactive Wastes

The following types of RWs enter the burial point for reliable neutralization and long-term storage:

- Solid radioactive waste (SRW).
- Liquid radioactive waste (LRW).
- Spent ionizing radiation sources.

The RWs that enter PBRWs are characterized by almost all known radionuclides, but the most commonly found include ^{137}Cs, ^{90}Sr, ^{60}Co, ^{3}H, ^{131}I, U, Pu, and Th.

Table 10.1. General characteristic of Radon SEs.

Name of enterprise	Departmental oversight	Volume of storehouses, m^3
Volgograd SE	Ministry of Construction of RF	1,800
Grozny SE	Ministry of Construction of RF	2,250
Murmansk SE	Ministry of Construction of RF	1,200
Nizhniy Novgorod SE	Ministry of Construction of RF	1,800
Novosibirsk SE	Ministry of Construction of RF	2,520
Kazan SE	Ministry of Construction of RF	1,200
Saratov SE	Ministry of Construction of RF	800
Irkutsk SE	Ministry of Construction of RF	2,000
Bashkotarstan SE	Ministry of Construction of RF	800
Khabarovsk SE	Ministry of Construction of RF	1,700
Chelyabinsk SE	Ministry of Construction of RF	2,600
Ekaterinburg SE	Ministry of Construction of RF	2,520
Samara SE	Ministry of Construction of RF	1,200
Rostov SE	Ministry of Construction of RF	650
Plant of Sosnovy Bor branch of SIA Khlopin Radium Institute	Ministry of Atomic Energy of RF	64,000
Moscow SIA Radon – incorporated ecologic-technological and scientific research center on RW utilization and security	Government of Moscow	68,700

Table 10.2. Characteristics of RWs.

Name of enterprise	Filled volume of storages, m^3	Total activity at enterprise, Bq
Volgograd SE	1,000	11.6×10^{14}
Grozny SE	1,250	0.2×10^{14}
Murmansk SE	400	1.7×10^{14}
Nizhniy Novgorod SE	1,000	2.1×10^{14}
Novosibirsk SE	850	3.0×10^{14}
Kazan SE	500	0.1×10^{14}
Saratov SE	400	1.9×10^{14}
Irkutsk SE	1,000	0.6×10^{14}
Bashkotarstan SE	500	0.7×10^{14}
Khabarovsk SE	900	2.2×10^{14}
Chelyabinsk SE	1,300	2.2×10^{14}
Ekaterinburg SE	1,250	1.2×10^{15}
Samara SE	300	0.3×10^{14}
Rostov SE	250	2.2×10^{14}
Leningrad SE	57,600	2.7×10^{15}
Moscow SIA Radon	61,000	2.2×10^{16}

The reprocessing and disposal of RWs and ionizing radiation sources results in the potential release of radioactive substances into the environment. These potential releases can be grouped into four zones:

- Zone of SRW burial.
- Zone of purified waters discharge.
- Zone of purified gases release.
- Zone of storage for spent sources.

Studies carried out at the Moscow SIA Radon have shown that concentrations of radioactive materials (RMs) in soil, vegetation, and water in the strict regime zone, sanitary-protection zone (SPZ), and observation zone are identical for all SE examined. Using the criteria of 95% confidence, the distinctions are insignificant. SEs operations have no influence on the environment. Their radiation state is influenced by externalities, and is not linked to the combines.

The RW characteristics of SEs of Russia are given in *Table 10.2*.

In accordance with the kind of RWs, their activity, and their method of RW technological reprocessing, the RWs are stored in special underground rectangular tanks made of overlapping, composite-reinforced concrete plates. The vessels for liquid wastes are made of reinforced concrete monolith and are lined with stainless steel. The RW burial technique in the near-surface layer is commonly accepted. Near-surface storehouses take the form of holes, trenches, superficial mines, and sumps. Barrow-type storehouses are also considered to be near-surface storehouses.

At the time of the Soviet disintegration, there were 35 PBRWs, with 16 of them on the territory of the RF, six on the territory of Ukraine, and a PBRW in each of the remaining republics. The general characteristics of RW storehouses of the RF are shown in *Table 10.3*.

Sources of the information for the technical and radioecological condition at the SEs are as follows:

- Annual accounting materials for regional administration of SEs.
- Logs of receipt of RWs.
- Descriptions of technical projects of SEs.
- Logs of the radiation control of environmental objects of SEs.
- Logs of the individual control of irradiation of SE personnel.
- Accounting materials on geographical and geomorphological inspection of the SEs regional location.

The RW depositories of the FSU that are now outside the territory of the RF are listed in *Table 10.4*. The total amount of RW kept in these depositories is estimated at 0.2×10^{16} Bq as of 1 January 1991.

10.3 Geoinformation System of Radioactive Waste Burial at the Special Enterprises of Russia

One conclusion that can be drawn from the analysis of the environmental protection activity in the sphere of the RW handling is that it is necessary to carry out work within the framework of the regional system of radioecological monitoring using advanced means of collecting and processing the significant data arrays. The information system consists of a database and multilayer cartographical application. The basic system modules can be used both in local computer networks and in remote access mode.

In developing the geoinformation system, a principle of its use as a regional system of tracking for the RW handling on territory of serving the SE is incorporated. The basic tools are as follows:

Table 10.3. General characteristics of RW storage facilities.

Name of enterprise	Capacity of storehouses, m^3	Filled volume of storehouses, m^3	Date of commis-sioning	Location of the object
Volgograd SE	1,800	1,000	1964	Gorodishchi district, 23 km from Volgograd
Grozny SE	2,250	1,250	1964	
Murmansk SE	1,200	400	1965	33 km northwest of Murmansk
Nizhniy Novgorod SE	1,800	1,000	1962	Semenov district, 80 km northeast of Nizhniy Novgorod
Novosibirsk SE	2,520	850	1968	18 km from Novosibirsk
Kazan SE	1,200	500	1964	9 km northeast of Kazan
Saratov SE	800	400	1964	Tatishchev district, 15 km from Saratov
Irkutsk SE	2,000	1,000	1964	35 km north of Irkutsk
Bashkotarstan SE	800	500	1964	Blagoveshchensk district, 20 km northwest of Ufa
Khabarovsk SE	1,700	900	1964	Western end of Khekhtsir mountain range, 40 km from Khabarovsk
Chelyabinsk SE	2,600,	1,300	1963	Sosnovsk district, 55 km from Chelyabinsk
Ekaterinburg SE	2,520	1,250	1964	Verkhhyaya Pyshma district
Samara SE	1,200	300	1963	Volga region, 35 km from Samara
Rostov SE	650	250	1962	Joint Aksai, Myasnikovo, Rodionov-Nesvetaev districts, 35 km from Rostov-on-Don
Leningrad SE	64,000	57,600	1962	At SPZ of Leningrad NPP
Moscow SIA Radon	68,700	61,000	1961	Sergiev-Posad district, 25 km north of Sergiev-Posad

Table 10.4. RW depositories in former Soviet countries other than the RF.

Object	Location and address
Kiev Special Combine	50 Krasnoznamennaya St., Kiev, Ukraine
Donetsk Special Combine	2nd Site, Adminposelok, Donetsk, Ukraine
Lvov Special Combine	4 Tajikskaya St., Lvov, Ukraine
Odessa Special Combine	Vladimira St., Odessa, Ukraine
Dnepropetrovsk Special Combine	24b Geroyev Stalingrada St., Dnepropetrovsk, Ukraine
Kharkov Special Combine	1 Sokolova St., Kharkov, Ukraine
Azerbaijan PBRW (Isotope Special Combine)	10 Izmira St., Baku, Azerbaijan
Yerevan Special Combine	47 Alaverdyan St., Yerevan, Armenia
Minsk SE for RW Management	Sosny Settlment, Zavodskoy District, Minsk, Belorussia
Georgia PBRW	2 Tolstoy St., Tbilisi, Georgia
PBRW at the Institute of Nuclear Physics of the Academy of Science of Kazakhstan	INP ASK, Almaty, Kazakhstan
Kyrgystan Special Combine	7a Fucik St., Bishkek, Kyrgystan
Latvia PBRW (Radon Enterprise)	11 Plata St., Riga, Latvia
Special Combine at the Institute of Physics of the Academy of Science of Lithuania	12 Gostauto St., Vilnius, Lithuania
Moldavia PBRW	210 Zavodskaya St., Kishinev, Moldavia
Tajikistan PBRW	Faizabad Settlement, Dushanbe, Tajikistan
Turkmenia Special Point for RW Disposal	57 Lakhuty St., Ashkhabad, Turkmenia
Uzbekistan PBRW	Institute of Nuclear Physics, Ulugbek Settlement, Tashkent, Uzbekistan
Estonia PBRW	Sillamyae Settlement, Estonia

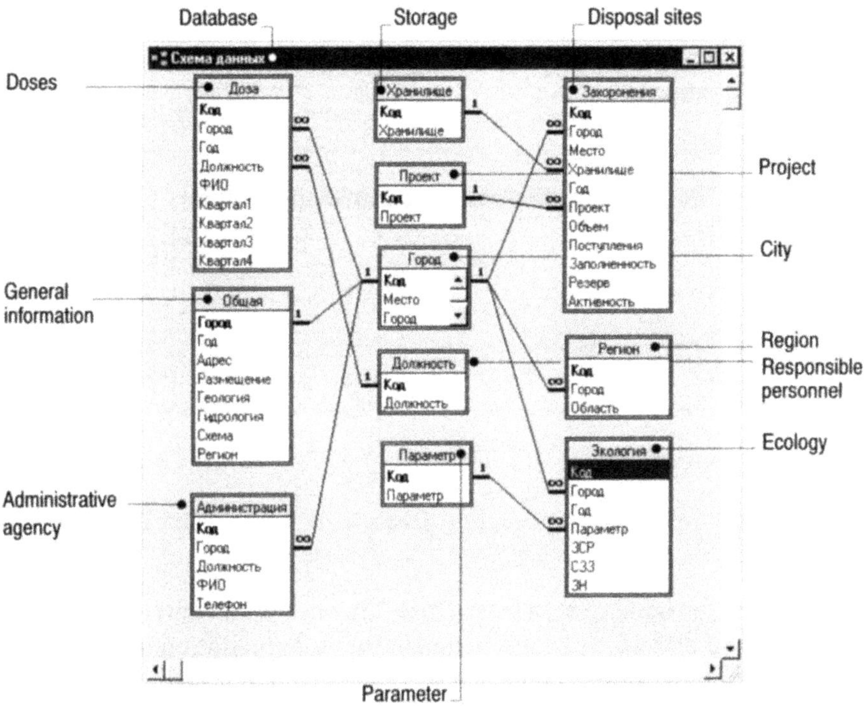

Figure 10.2. Data scheme of RW handling system.

- Microsoft Windows environment.
- DBMS Microsoft Access application.
- DDE (Dynamic Data Exchange).
- OLE (Object Linking and Embedding).

A conceptual model can be developed on the basis of an analysis of the activity of the regional enterprise on RW burial. The data scheme and links between them are given in *Figure 10.2*. The group of data, incorporated under "General" includes such fields as "Address," "Scheme," and "Region." Using these attributes, it is possible to introduce these objects into the computer cartography software. This enables complex monitoring of the radiation situation to be performed for RW burial zones, the SE environment, and the objects of a potential radioactive hazard, all within the framework of one database. The submitted data scheme includes not only information blocks linked to the RW burial technology, but also new modules relating to radiation and radioecological monitoring. The modules

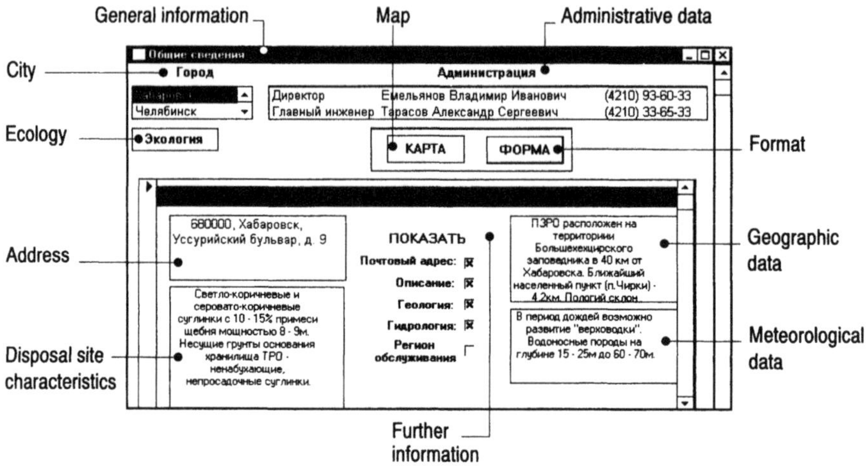

Figure 10.3. "General Information" form.

can be given designations such as "Dose" (to ensure the individual control of the personnel) and "Ecology" (to monitor radionuclide content of strict mode, SPZs, and observation zones at RW burial facilities).

The interface to the information is realized on the basis of multiple queries to the forms, which allow the information to be displayed in both a textual and in a graphic form. The "General Information" form is indicated in *Figure 10.3.* The analysis of parameters of the radiation monitoring is affected by the complex composite forms. They are presented in *Figure 10.4.*

The input and analysis of data related to cartography is performed with the help of the geographical information system (GIS). The GIS uses sets of graphic files with images of maps and/or plans of regions. The set is incorporated by a general coordinate system. Grids of benchmark nodes are applied in order to account for possible distortions of the image of the region. The system provides for setting of a point on a map or plan, data input with binding to a place, display of information layers (on the screen and printer), construction of the isolines of studied value, etc.

The information layer "PBRW of Russia" includes the scheme of serving zones of 16 enterprises, plans of enterprise industrial zones, and layouts of vicinities. The complete information about PBRW is stored in the database together with the data concerning radioecological monitoring of the industrial zone and its vicinity. An example of visualizing the

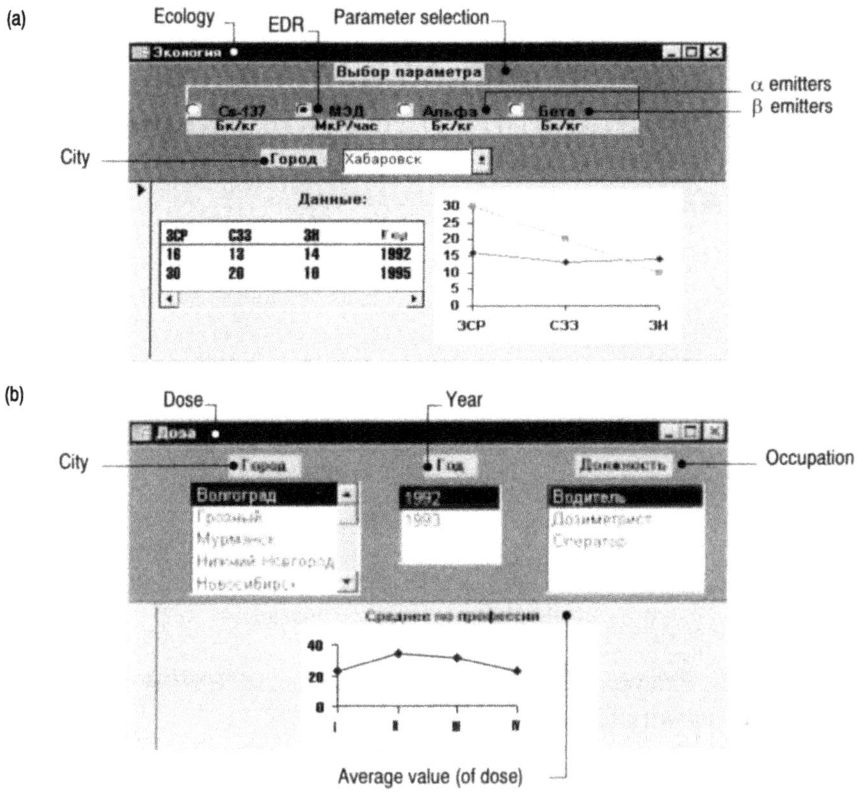

Figure 10.4. "Ecology" (A) and "Dose" (B) forms.

information with the GIS application is presented in *Figure 10.5*. The main blocks of the developed geoinformation complex were already in use for information maintenance of the international Radleg project.

The geoinformation complex also includes such important information as mathematical models of radionuclide migration from PBRW storages. The study of the techniques of RW processing and burial has shown the necessity of developing the mathematical model. The above-mentioned information system allows long-term forecasts to be performed for environmental conditions by mathematical modeling of processes of radionuclide migration from a burial zone. One can evaluate these processes to detect potential hazardous tendencies and possible emergencies. Use of the given model provides an opportunity for predicting the migration of the radioactive substances from the places of RW burial. The model

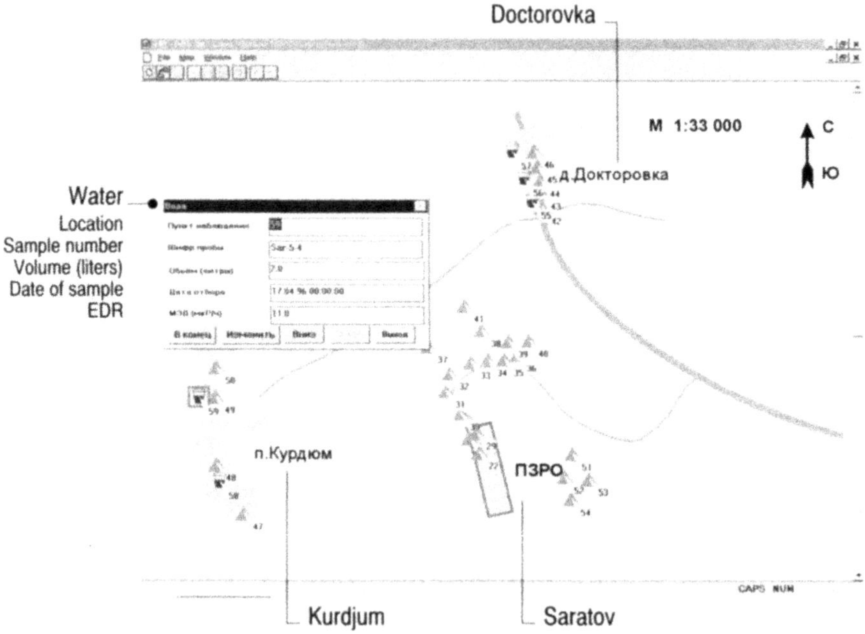

Figure 10.5. Examples of GIS data visualization of "PBRW of Russia": Monitoring network of the Saratov SE.

allows forecasts to be carried out that facilitate the choice of burial sites, the optimum activity levels for RWs sent to burial, and the most effective technologies of RW processing prior to burial.

Thus, the geoinformation system developed allows not only the study of efficient radiation monitoring control on the SE serving the regional territory of RW burial, but also a prediction of the radiation situation.

Information flows on all kinds of SE activity, as well as the radio-ecological monitoring data of environmental objects, such as on combines (enterprises) and on Moscow and the Moscow region, are concentrated in the Center of Information Processes and Technologies. There, databases are formed, mathematical processing and analysis of the information is conducted, and forecasts of the developing radiation situation are made. From here the processed information is sent to the interested organizations for applicable decision making: the Government of Moscow, the prefectures of administrative districts, the State Sanitary and Atomic Supervision units, the divisions on nature protection, and staff of civil defense.

Bibliography

Energoatomizdat, 1993, *Life and Radiation* (in Russian), Energoatomizdat, Moscow, Russia (originally published in 1990 by National Council on Radiation Protection of Great Britain, London, UK).

Goskomstroy of the USSR, 1985, *Sites for Reducing Toxicity and Disposal of Toxic Industrial Wastes: The Fundamentals of Designing*, SNiP 2.01728-85, Goskomstroy of the USSR, Moscow, Russia.

Minzdrav of the USSR, 1986, *Sanitary Rules of Radioactive Wastes Handling (SPORO–85)*, Minzdrav of the USSR, Moscow, Russia.

Minzdrav of the USSR, 1989, *Collection of Standards and Rules on Radiation Safety in Nuclear Power Engineering*, Vol. 1, Minzdrav of the USSR, Moscow, Russia.

RRC-KI (Russian Research Center – Kurchatov Institute), 1992, *The Assessment of the Practicability of the Construction Program of Modern Storage Facilities for Low-Level Waste Disposal in the Territory of the Russian Federation*, RRC-KI, Moscow, Russia.

Serebryakov, B.E., 1995, The calculation of radionuclide migration from the burial places, *Atomic Energy*, **79**(5):381–386.

SIA (Scientific Industrial Association) Radon, 1995, *The Russian Specialized Enterprises on Radioactive Waste Handling: Radioecological Aspects*, SIA Radon, Moscow, Russia.

Sobolev, I.A., and Khomchik, L.M., 1983, *Rendering the Neutralized Radioactive Wastes at Centralized Points*, Energoatomizdat, Moscow, Russia.

Sobolev, I.A., Korenkov, I.P., Khomchik, L.M., and Prokazova, L.M., 1989, *Environmental Protection During the Process to Reduce Radioactive Waste Toxicity*, Energoatomizdat, Moscow, Russia.

Sobolev, I.A., 1993, *The Radioactive Waste Handling of Research and Industrial Centers*, Government of Moscow, Department of Engineering Service, SIA Radon, Moscow, Russia.

Sobolev, A.I., Tikhomirov, V.A., and Zubov, V.Yu., 1996, Computing regional information for a complex radioecological monitoring system, *Proceedings of the Second International Scientific and Practical Conference "The Problems of Quality Control of Environment of Cities,"* 1–4 October, Moscow, Russia.

Timofeev, E.M., Ozhovan, M.I., Kachalov, M.B., Arustamov, A.Eh., and Shiryaev, V.V., 1995, Rendering of radionuclide sources at the regional disposals of Russian Federation, *Proceedings of the First International Scientific Practical Conference "Radioecological Safety of the Modern Civilization: Social and Cultural Approaches, Information Technologies, Economical Structures,"* 6–8 June, Moscow, Russia.

TsITP (Central Institute for Standardization of Designs), 1990, *The Construction Catalogue: Standard Projects of Enterprises, Buildings, and Structures*, TsITP, Moscow, Russia.

VNPO (Scientific and Industrial Association) Soyuzpromgaz, 1990, *Technical and Economical Investigations on the Creation of a Storage Facility for RW Disposal of Intermediate and Low Specific Activity Level*, VNPO Soyuzpromgaz, Moscow, Russia.

Chapter 11

Conclusion

Nikolai N. Egorov

This report provides a general idea of the scale and complexity of the problems of the radiation legacy of the former Soviet Union (FSU). Analysis of the material shows that the organizations participating in this project possess large volumes of data in particular sectors of the investigations. However, some data remain secret. This has led to incomplete descriptions in some chapters.

The reviews on nuclear power plants, marine nuclear power installations and their repair and maintenance facilities, sea dumping of radioactive wastes (RW), and radiation consequences of nuclear weapon tests meet the overview's objectives to the extent possible. At the same time, the data on nuclear fuel-cycle enterprises, research nuclear reactors, and geological disposal of RWs and accompanying problems are less complete.

Among the overview's general drawbacks are the following:

- Nonuniformity of data presentations, which, in the ever-changing situations at the sources of radiation hazard, complicates the estimation of the risks to the biosphere at any given moment.
- Data, which are necessary to achieve the project's goals, and which are presently inaccessible, have not been explicitly identified.

- The difficulties in getting this data in the future have also not been identified.
- The partial results of the inventory carried out in Russia at sites of storage and disposal of RWs.

These and other disadvantages, to some extent, are a consequence of objective causes conditioned by a huge amount of information, inaccessibility of some important data, lack of experience with large flows of information, and limited financial resources.

Nevertheless, this overview makes it possible to identify, on the whole, the main potentially hazardous objects in Russia. These following objects may be, in emergency situations (earthquakes, floods, etc.), the main sources of radiation contamination of the biosphere:

- Objects situated in northern and Pacific regions of Russia (Murmansk and Arkhangelsk regions, Primorsk land) that are connected with operation and utilization of Navy nuclear submarines and civilian nuclear-powered ships.
- Some objects of the Industrial Association Mayak in Chelyabinsk region (Karachay Lake and the cascade of water reservoirs).
- The surface water reservoirs for storage of liquid radioactive wastes (LRWs) at the Siberian Chemical Combine (SCC) in Tomsk region and at the Mining and Chemical Combine (MCC) in Krasnoyarsk land.
- The repositories of spent nuclear fuel and RWs at nuclear power plants (NPPs), particularly those with RBMK reactors.
- The repositories of RWs at special Radon-type enterprises in different regions of Russia.

These objects are the subject of public anxiety in the above-mentioned regions of Russia.

The underground disposal of LRWs at the SCC in the Tomsk region and at the MCC in Krasnoyarsk land is controversial. However, according to conclusions of experts of the Ministry of the Russian Federation for Atomic Energy (Minatom), as well as a number of institutions of the Russian Academy of Sciences and other departments, they are not hazardous to the environment and the population. However, RWs in underground disposal sites contain long-lived radionuclides, and many hundreds of years are needed for their full decay. For this reason, long-term, thorough control of the disposal sites is necessary, and must include additional investigations. It should be noted that the possibility of deep well injection of RW disposal is rather limited because it requires suitable geological structures.

In the framework of further evaluation of the radiation legacy of the FSU, it would be worthwhile to:

- Continue to improve and verify data in each sector of investigation through the organizations participating in the project with stage-by-stage exploration of sites and objects of particular concern.
- Provide access to more accurate data and introduce previously inaccessible data as soon as they are available.
- Carry out ranking of risks, on both local and global scale, for two or three of the most potentially hazardous objects.
- Evaluate measures to lower risk including preliminary cost analysis.
- Connect the available database with a geographical information system (GIS).

To deal with these tasks, it is advisable to continue cooperation with the International Institute for Applied Systems Analysis' (IIASA's) Radiation Safety of the Biosphere project and the International Science and Technology Center (ISTC) Radleg project. The Radleg project aims to develop a complete information system to evaluate the radiation legacy of the FSU. This database should help set priorities for remediation of the contaminated areas, and, thus, prevent further contamination of the biosphere.

On the whole, the analytical overview is the first important stage in the evaluation of the radiation legacy of the FSU. It provides information for understanding the severity of present and potential problems of the radiation legacy, and it can serve as a starting point for further evaluation of methods for eliminating these problems.

Appendix A: List of Abbreviations and Acronyms[1]

AD	Admissible discharge – activity of a radionuclide discharged with wastewaters, which satisfies the condition that the total exposure dose due to all radionuclides present in the mixture does not exceed 0.05 mSv
AD	Autonomous district (in *Table 9.11*)
AECC	Angarsk Electrolysis Chemical Combine
AMB	Channel-type, uranium-graphite, water-cooled reactors with tabular fuel elements
AMCA	Argun Industrial Mining and Chemical Association
AO	Joint Stock Society
AR	Admissible release – activity of a radionuclide release in gas and aerosol states, which satisfies the standard applied in nuclear power engineering of not exceeding the population equivalent exposure dose (equal to 0.2 mSv)
AUDR	All-Union Distributed Registry
CHMP	Chemico-hydrometallurgical plant
CIS	Commonwealth of Independent States
CNPP	Chernobyl nuclear power plant
CTF	Critical test facility
DBMS	Database management system
DVZ	Far-East plant
ECP	Electrochemical plant
EDR	Exposure dose rate
EGP	Channel-type, uranium-graphite, water-cooled reactors with tabular fuel elements
EURT	East-Urals radioactive trace
FA	Fuel assembly
FE	Fuel element
FEI	Institute of Physics and Power Engineering
FM	Fissile material
FMS	Fundamental and methodical study
FP	Fission product
GEOKHI	Geochemical Institute
GIPE	State Institute for Applied Ecology
GIS	Geographical information system
GKHK	see MCC
GMP	State Machine-building Enterprise
GREC	Gypsum and rare earth concentrate

[1]Many entries in this list are derived from transliterations of Russian acronyms and abbreviations.

HEU	Highly enriched uranium
HMP	Hydrometallurgical process
HOB	Height of burst
HPP	Heating and power plant
IA	Industrial Association
IASAP	International Arctic Sea Assessment Project
IBRAE	Institute for Safety of Nuclear Power
IC	Ionization chamber
IE	Industrial underground peaceful nuclear explosion
IER	Institute of Experimental Reactors
IEXR	Ion exchange resin
IGEM	Institute of Geology of Ore Deposits, Petrography, Mineralogy, and Geochemistry
IGKE	Institute of Global Climate and Ecology
IIASA	International Institute for Applied Systems Analysis
IMP	Institute of Molecular Physics
INES	International Nuclear Events Scale
INP	Institute of Nuclear Physics
INR	Institute of Nuclear Reactors
INTEST	Minatom Sociological Association
IR	Induced radionuclide
ISTC	International Science and Technology Center
ITEF	Institute of Theoretical and Experimental Physics
LEU	Low enriched uranium
LLN	Long-lived nuclide
LRW	Liquid radioactive waste
MAC	Mitigation of accident consequences
MCC	Mining and Chemical Combine
MD	Mining department
ME	Military exercise with a nuclear detonation
MIFI	Moscow Engineering and Physics Institute
MMS	Murmansk Marine Steamship
MNPO	Moscow Scientific and Industrial Association
MOX-fuel	Mixed (U, Pu) oxide fuel
MPD	Maximum permissible discharge – radionuclide activity discharged with wastewaters, where the total radiation dose caused by all radionuclides present in the mixture does not exceed dose limits established for the critical organs' groups
MPR	Maximum permissible release – scientific and technical specification of release rate, determined on the basis of sanitary standards for dose limits, taking into account all pathways of external and internal irradiation
MSU	Moscow State University

MTR	Missile testing range
ND	Normative document
NFC	Nuclear fuel cycle
NIIAR	Research Institute of Atomic Reactors
NIIP	Research Instrumentation Institute
NIKIET	Research and Design Institute of Power Engineering
NPB	Radiation Safety Standard
NPO	Research and Industrial Association
NPP	Nuclear power plant
NRG	Noble radioactive gas
NRM	Norms of Radiation Safety
NRN	Natural radionuclide
NS	Nuclear submarine
NTSNZ	Northern test site Novaya Zemlya
NVNPP	Novovoronezh nuclear power plant
NWR	Nuclear weapon-related test
OPB	General Safety Rules
OSP	Basic Sanitary Rules
PBRW	Point of burial of radioactive wastes
PC_B	Permissible concentration (level) of volumetric activity of a radionuclide in atmospheric air or in drinking water (ration)
PIC	Public information center
PNE	Peaceful nuclear explosion
PO	Production amalgamation
R&D	Research and development
RAD project	Radiation Safety of the Biosphere project
RAMS	Russian Academy of Medical Science
RAS	Russian Academy of Sciences
RBMK	Channel-type, uranium-graphite, water-cooled power reactor
RFNC-VNIIEF	Russian Federal Nuclear Center – VNIIEF (All-Russia Research Institute of Experimental Physics)
RI	Minatom Radiochemistry Institute
RM	Radioactive materials
RRC-KI	Russian Research Center – Kurchatov Institute
RSMDR	Russian State Medical and Dosimetric Registry
RT-1,2	Radiochemical plants for spent nuclear fuel reprocessing
RTE	Russian Transport Enterprise
RW	Radioactive wastes – solutions, products, materials, and biological objects that cannot be used further and that contain amounts of radioactive substances that exceed values established in acting rules and standards
SAM	Study of accidental models and emergencies

SCC	Siberian Chemical Combine
SCR	Self-sustained chain reaction
SDBMT	Special Design Bureau of Marine Technology
SE	Special enterprise
SKHK	see SCC
SNF	Spent nuclear fuel
SPORO	Sanitary Rules for Radioactive Waste Management
SPR	Ship propulsion reactor
SPZ	Sanitary-protection zone
SRC	State Research Center
SRW	Solid radioactive waste
SRZ	Ship Repair Plant
STS	Semipalatinsk test site
SZ	Surveillance zone
TIC	Test of industrial nuclear charge for use in peaceful activities
TK	Process channel
TNT	Trinitrotoluene
TR	Transuranium radionuclide
TSEMI	Central Institute for Economics and Mathematics
TsITP	Central Institute for Standardization of Designs
UECC	Urals Electrochemical Combine
UL	Underground leaching
VNIICHT	All-Russia Research Institute of Chemical Technology
VNIIAES	All-Russia Research Institute for Atomic Power Plants
VNIIEF	All-Russia Research Institute for Experimental Physics
VNIINM	All-Russia Research Institute for Non-Organic Materials
VNIITF	All-Russia Research Institute for Technical Physics
VNIPIPT	All-Russia Research and Design Institute of Production Engineering
VNPO	Scientific and Industrial Association
VVER	Water-moderated, water-cooled power reactor
WGP	Weapon grade plutonium
WIE	Weapons effect test

Appendix B: Units of Measurement

Radionuclides' activity and radiation dose values, as a rule, are presented in the overview in the International System of units:

Becquerel (Bq)	A unit of radionuclide activity in a radiation source; 1 Bq corresponds to 1 decay per second for any radionuclide.
Gray (Gy)	A unit of absorbed dose; 1 Gy corresponds to 1 J/kg.
Sievert (Sv)	A unit of equivalent dose; for x-ray and gamma- and beta-radiation, 1 Sv corresponds to an absorbed dose of 1 J/kg.
Nonsystem units	Used in the overview either in parallel or in cases when official sources are cited or primary measurements' data taken from original publications are presented.
Curie (Ci)	A non-SI unit of radionuclide activity in a radiation source; 1 Ci corresponds to 3.7×10^{10}Bq.
Roentgen (R)	A non-SI unit of exposure dose; 1 R is that amount of gamma-radiation required to produce ions carrying 2.58×10^{-4} C of electric charge per kg of dry air under normal conditions. 1 R would approximately correspond to absorbed dose in a biological soft tissue equal to 9.3×10^{-3} Gy (on average).
Rem	A non-SI unit of equivalent dose (a biological equivalent of Roentgen); 1 rem corresponds to 0.01 Sv.

To form multiple units the following prefixes are used:

Prefix	Numeric value
micro (μ)	10^{-6}
milli (m)	10^{-3}
centi (c)	10^{-2}
kilo (k)	10^{3}
mega (M)	10^{6}
giga (G)	10^{9}
tera (T)	10^{12}
peta (P)	10^{15}
exa (E)	10^{18}

Index

For Product Safety Concerns and Information please contact our EU representative GPSR@taylorandfrancis.com
Taylor & Francis Verlag GmbH, Kaufingerstraße 24, 80331 München, Germany

www.ingramcontent.com/pod-product-compliance
Ingram Content Group UK Ltd.
Pitfield, Milton Keynes, MK11 3LW, UK
UKHW022005180926
475701UK00003B/208
* 9 7 8 1 1 3 8 9 8 4 4 6 2 *